Fang Wang

Shape Optimization for Piezoceramics

Fang Wang

Shape Optimization for Piezoceramics

VDM Verlag Dr. Müller

Imprint

Bibliographic information by the German National Library: The German National Library lists this publication at the German National Bibliography; detailed bibliographic information is available on the Internet at http://dnb.d-nb.de.

Cover image: www.purestockx.com

Publisher:
VDM Verlag Dr. Müller Aktiengesellschaft & Co. KG, Dudweiler Landstr. 125 a, 66123 Saarbrücken, Germany,
Phone +49 681 9100-698, Fax +49 681 9100-988,
Email: info@vdm-verlag.de

Produced in USA and UK by:
Lightning Source Inc., La Vergne, Tennessee, USA
Lightning Source UK Ltd., Milton Keynes, UK
BookSurge LLC, 5341 Dorchester Road, Suite 16, North Charleston, SC 29418, USA

ISBN: 978-3-639-01247-7

Acknowledgements

I would like first to express my deep appreciation and gratitude to Prof. Dr. Michael Dellnitz, my academic advisor, for his invaluable supervision, encouragement and continuous support throughout this work. He opened the door to this intriguing research area for me.

I am grateful to my co-advisor, Prof. Dr.-Ing. Jörg Wallaschek, for his comments and suggestions in this work. My special thanks are conveyed to Prof. Dr. Oliver Junge at Technische Universtät München and Dr. Robert Preis for their encouragement and for providing me with helpful suggestions. Thanks also goes to Dr.-Ing. Tobias Hemsel for his careful reading and comments.

My gratitude goes to my colleagues at Chair of Applied Mathematics during my studies in Paderborn. I am greatly thankful to Mirko Hessel-von Molo for valuable discussions, suggestions and comments, Dr. Oliver Schütze and Stefan Sertl for technique support, and Marcus Post for many stimulating discussions during my work. I appreciate to Bianca Thiere, Dr. Kathrin Padberg, Sina Ober-Blöbaum, Katrin Witting, Marianne Kalle, Arvind Krishnamurthy and Alessandro Dell'Aere for their encouragement, friendship and help.

I acknowledge Deutsche Forschungsgemeinschaft (DFG), the Graduiertenkolleg of the Paderborn Institute for Scientific Computation (PaSCo) and Deutscher Akademischer Austausch Dienst (DAAD) for providing me with scholarships which helped me to finish my dissertation research. Thanks also goes to all the members of the Graduiertenkolleg of PaSCo for enlightening discussions during our workshops on Application-oriented Modeling and Development of Algorithms, and friendship, in particular to Dr. Valentina Damerow, Dr. Bo Fu, Dr. Martin Ziegler and Nicolai Neumann.

Furthermore, I would like to thank Prof. Eusebius Doedel at Concordia University for his extremely helpful assistance with the use of AUTO2000.

Mostly I would like to thank my parents in China for their love and support throughout my life, and Dr. Xiaoping Jia at Tsinghua University for his constant patience, understanding and support.

Contents

Abstract   **ix**

Zusammenfassung   **xi**

1   Introduction   **1**

   1.1   Shape Optimization   1

   1.2   An Overview of *Shape Optimization* in Piezoelectric Actuator Design . .   3

   1.3   Motivation of this Work   6

   1.4   Organization of this Dissertation   12

2   Piezoelectricity   **15**

   2.1   Piezoelectric Effect   15

       2.1.1   Introduction   15

       2.1.2   Piezoelectric Constitutive Equations   17

   2.2   Piezoelectric Materials   18

   2.3   Piezoelectric Actuator   22

3   Derivation of General Eigenfunctions and Equations of Motion   **25**

   3.1   A Nonlinear Model   26

   3.2   General Eigenfunctions Derivation for Piezoceramics with a Linear Model   29

       3.2.1   A Linear Model   30

       3.2.2   General Eigenfunctions Derivation for Different Geometries of

               Piezoceramics   33

       3.2.3   Numerical Solutions   38

3.3 Derivation of the Equation of Motion for a General Shape 38

4 Solutions of Equations of Motion and Nonlinear Dynamical Model Analysis 47

4.1 Numerical Solutions for a Linear Model 47

4.2 Numerical Solutions for a Nonlinear Model 51

4.3 Nonlinear Effects Analysis 54

5 Multi-Objective Optimization Problems (MOOPs) 57

5.1 A Brief Introduction to MOOPs 57

  5.1.1 Formulation of the Multi-objective Optimization Problem 58

  5.1.2 Pareto Optimality and Pareto Set 58

5.2 Methods to Solve MOOPs 59

  5.2.1 An Overview 59

  5.2.2 Software 60

5.3 A Set Oriented Multilevel Subdivision Technique 60

  5.3.1 A Set Oriented Multilevel Subdivision Technique 60

  5.3.2 Software 64

5.4 A GAIO Model for the Multi-Objective Optimization Problem 64

  5.4.1 MOOPs in Piezoelectric Actuator Design 64

  5.4.2 The Multi-objective Shape Optimization Problem 65

  5.4.3 A GAIO Model for the Optimization Problem 73

6 Results and Discussion 75

6.1 Optimization Results 75

  6.1.1 Multi-Objective Optimization Solutions for One Parameter (quadratic curves) 75

  6.1.2 Multi-Objective Optimization Solutions for Two Parameters (cubic B-spline curves) 79

6.2 Discussion 81

7 Summary and Outlook 87

7.1 Summary 87

7.2 Outlook 89

A   COMSOL Multiphysics (former FEMLAB)   91

B   A Brief Introduction to AUTO2000   95

Bibliography   97

List of Symbols   103

List of Tables   107

List of Figures   109

CONTENTS

Abstract

Piezoelectric actuators are being used increasingly in various novel applications. One of piezoelectric actuator design goals is to improve its performance for a certain mass of piezoelectric materials. Shape optimization is one important way to improve the performance of a piezo by changing its geometry. However, academic and industrial research of shape optimization is still developing, especially with several objectives considered simultaneously. This dissertation focuses on numerical modeling and multi-objective optimization of the shape of piezoceramics.

This work first explores the development of mathematical models and their related modeling procedure in detail. A mathematical model is introduced to describe the property of a piezo excited in resonance under weak electric fields. General eigenfunctions for piezoceramics with different shapes are derived. The description and numerical computation of a boundary value problem and the nonlinear dynamical behavior analysis are also presented. Both usual shapes (e.g. a rectangular shape) and (compared to the rectangular shape) unusual ones (e.g. a shape with curved sides) are considered, and the results show that some curved side piezoceramics perform better than those with a rectangular shape using both linear and nonlinear models for the dynamics.

In the next step a multi-objective shape optimization problem for the design of piezoelectric actuators is introduced. Two objectives, maximum amplitude (better performance) and minimum curvature (simple manufacturing), need to be optimized at the same time. The optimization is conducted with a subdivision algorithm based on the software package GAIO and the corresponding Pareto-optimal solutions are obtained both for linear and nonlinear models. The results show that there is indeed an advantage in using more complex shapes, as the Pareto set obtained using two design variables (in this case parameterizing a cubic B-spline) has substantially better objective function values than one with one design variable (in this case a quadratic curve).

Key words

Piezoceramics; Shape optimization, Eigenfunction; Boundary value problem; Bifurcation; Piezoelectric actuator; Multi-objective optimization; Subdivision algorithm

Zusammenfassung

Piezoelektrische Aktuatoren finden immer häufiger in neuartigen Produkten verschiedenster Art Anwendung. Eines der Ziele bei der Entwicklung piezoelektrischer Aktuatoren ist die Optimierung der Leistung für eine bestimmte Menge piezoelektrischen Materials. Die Formoptimierung stellt eine wichtige Möglichkeit zur Verbesserung der Leistung eines Piezobauteils durch Änderung seiner geometrischen Eigenschaften dar. Sowohl die akademische wie auch die industrielle Untersuchung der Formoptimierung befinden sich jedoch noch in der Entwicklung, insbesondere für den Fall mehrerer, gleichzeitig betrachteter Zielfunktionen. Diese Dissertation beschäftigt sich mit der numerischen Modellierung und Mehrzieloptimierung der Form piezokeramischer Bauteile.

In dieser Arbeit werden zunächst die Entwicklung mathematischer Modelle und der dazugehörigen Modellierungsverfahren detailliert betrachtet. Es wird ein mathematisches Modell eingeführt, das die Eigenschaften eines piezoelektrischen Bauteils in Resonanz mit einem schwachen elektrischen Wechselfeld beschreibt. Für dieses Modell werden Eigenfunktionen in Abhängigkeit von der Form des Bauteils hergeleitet. Weiterhin werden die numerische Behandlung des sich ergebenden Randwertproblems und die Analyse des nichtlinearen dynamischen Verhaltens vorgestellt. Dazu werden sowohl gewöhnliche Formen (wie zum Beispiel Quader) als auch (im Vergleich dazu) ungewöhnliche Formen (z. B. mit gekrümmten Seitenflächen) betrachtet. Die Ergebnisse zeigen, dass einige gekrümmte Bauteile eine bessere Leistung als vergleichbare quaderförmige Teile zeigen, sowohl unter Verwendung des linearen wie auch unter Verwendung des nichtlinearen Modells.

Im nächsten Schritt wird das Mehrziel-Form-Optimierungsproblem für die Gestaltung piezoelektrischer Aktuatoren eingeführt. Zwei Zielfunktionen, die zu maximierende Amplitude (höhere Leistung) und die zu minimierende Krümmung (einfachere Herstellung) sollen dabei gleichzeitig optimiert werden. Die Optimierung wird mit einem im Softwarepaket GAIO implementierten Unterteilungsalgorithmus durchgeführt, wobei Paretooptimale Lösungen sowohl für das lineare wie auch für das nichtlineare Modell bestimmt werden. Den Ergebnissen entnimmt man, dass komplexere Formen vorteilhaft sind, da die Paretomenge für den Fall zweier Design-Variablen (die hier von der Parametrisierung eines kubischen B-Splines herrühren) auf deutlich bessere Zielfunktionswerte führt als für den Fall einer Variablen (die eine parabolische Form parametrisiert).

Chapter 1

Introduction

This chapter consists of four sections. In the first section the definition and classification of shape optimization are presented briefly. The state of the art in shape optimization of piezoelectric actuator based on the shape optimization classification is presented in the second section. The motivation of this work is proposed and its feasibility is analyzed in the third section. The fourth section presents the organization of this dissertation.

1.1 Shape Optimization

Former research into the optimization of structures has been attempted since antiquity; the designer would choose the shape and materials for the construction using intuition and experience. Since ancient times this technique has proved effective, and for centuries engineering landmarks such as castles, cathedrals, and ships were all built without mathematical or mechanical theories. However, from the time of Galileo and Hooke, engineers and mathematicians have developed theories to determine stress, deflections, currents and temperature inside structures. Evidence of structural optimization in the modern era was first documented in the 17th century by Galileo in his treatise, where the optimal shape of beams was investigated.

Structural optimization is a major concern in the design of mechanical systems in the industry (civil engineering, auto manufacturing, aeronautics, aerospace). In the past few decades, it has become possible to turn the design process into algorithms thanks to ad-

1

vances in computer technology. The incresing modern trend is to use numerical software which is used to analyze and optimize many possible designs simultaneously, making optimal design an automatic process. By applying different computer-oriented methods, the topology and shape of structures can be optimized and hence designs are systematically improved. These possibilities have stimulated an interest in the mathematical foundations of structural optimization [Che00].

Definition. *Shape optimization* [1] usually has a very broad meaning. It can be viewed as a part of the important branch of computational mechanics called structural optimization. In structural optimization problems one tries to set some data of the mathematical model that describe the behavior of a structure, therefore one can find a situation in which the structure exhibits a priori of given properties. In *shape optimization*, as the term indicates, optimization of the geometry is of primary interest.

Classification. From our daily experience we know that the efficiency and reliability of manufactured products depend on geometrical aspects, among others. Therefore, it is not surprising that optimal shape design problems have attracted the interest of many applied mathematicians and engineers.

Nowadays *shape optimization* represents a vast scientific discipline involving all problems in which the geometry (in a broad sense) is subject to optimization. It is indispensable in the design and construction of industrial structures. For example, aircraft and spacecraft have to satisfy, at the same time, very strict criteria on mechanical performance while weighing as little as possible. The shape optimization problem for such a structure consists in finding a geometry of the structure which minimizes a given function (e.g. such as the weight of the structure) and yet simultaneously satisfies specific constraints (like thickness, strain energy, or displacement bounds).

In some publications the term **shape optimization**[2] is used in a comparatively restrict sense. As it is also called geometric optimization, its emphasis is not on changing size or topology but geometry. Haslinger considered shape optimization in a restricted sense as

---

[1] *Shape optimization* (italic) will be used to represent the term in a broad sense.

[2] Shape optimization (roman) will be used to represent the term in a more restricted sense.

a branch of *shape optimization* in a broad sense. For a finer classification, three branches of *shape optimization* are distinguished as follows [HM03]:

1. **size optimization**: a typical size of a structure, such as the thickness of a shell or the radius of a circular stress element, is optimized. This class of problems has been under modern investigation for decades.

2. **shape optimization** (also called geometric optimization): the shape of a structure is optimized without changing the topology. Shape optimization is a classical field of the calculus of variations, optimal control theory and structural optimization. Due to its increased difficulty relative to size optimization, the geometrical changes have historically been limited; however, it has gained importance in the aircraft and automotive industries, as well as others, providing improvements to turbines, airfoil shapes and connecting arms. Size optimization is a subset of shape optimization.

3. **topology optimization**: the topology of a structure, as well as the shape is optimized by, for example, creating holes.

In this work we only focus on (2). One important feature of shape optimization is its *interdisciplinary character*. First, the problem has to be posed from a mechanical point of view. Then one has to find an appropriate mathematical model that can be used for the numerical realization. In this stage no less than three mathematical disciplines interfere: the theory of partial differential equations (PDEs), approximation of PDEs (usually by finite element methods), and the theory of nonlinear mathematical programming.

1.2   An Overview of *Shape Optimization* in Piezoelectric Actuator Design

Piezoelectric actuators are being increasingly used in various novel applications. A piezoelectric actuator usually consists of two main components: an electric part, which is the piezoelectric material block, can convert electrical energy into mechanical energy, and a mechanical part, which is a flexible structure, can convert and amplify the output displacement in the desired direction and magnitude [LXKS01]. One of the important issues of using piezoelectric actuators is to improve their performance for a certain mass of piezoelectric material, which is the goal of piezoelectric actuator design.

Design of piezoelectric actuators has been greatly advanced during the past ten years. Usually the performance of a piezoelectric actuator can be improved by optimizing the mechanical part, the electric part, or both. In the previous work, the topology and shape of the mechanical part were designed, however, the location and the shape of the piezo-electric material were fixed.

The design of the electric part has also been developed in recent years. In this section, the state of the art of *shape optimization* of piezoelectric materials is reviewed according to the classification in Section 1.1, together with the methods applied for solving the different optimization problems.

Sizing optimization.   Sizing optimization, also called cross-sectional optimization, has already been thoroughly studied. Main et al. [MGH94] optimized both the placement and size of the piezoceramics (PZT) in beams and plates. Jiang et al. [JNL00] stud-ied the physical parameters of the PZT plates and found that large areas of PZT plates with a small width are conducive to attaining higher velocities of actuators. Von Wagner ([vWH02], [vWH03]) discussed the influence of the radius of a piezo rod on the vibra-tion amplitude and analyzed the nonlinear property under weak electricity. Fu [Fu05] solved a constrained two-objective optimization problem involving continuous (dimen-sions of a piezoelectric transducer) and discrete design variables (material types) by using an elitist non-dominated sorting genetic algorithm (NSGA-II). The two objectives are the maximum vibration amplitude and the minimum electrical input power. Heikkola et al. [HMN05] optimized three objectives by considering the length of the head mass and the radius of the tip of an ultrasonic transducer with NIMBUS (Nondifferentiable Interactive Multiobjective BUndle-based optimization System).

Shape optimization.   Shape optimization is a new topic in this area. Several publi-cations are recommended for obtaining a general comprehension of shape optimization ([HM03], [SZ92], [Pra74], [Ric95]). Haslinger and Mäkinen [HM03] treated sizing and shape optimization in a comprehensive way, covering everything from mathematical theory (existence analysis, discretizations, and convergence analysis for discretized problems) through computational aspects (sensitivity analysis, numerical minimization methods) to industrial applications. Some of the applications included are contact stress minimization for elasto-plastic bodies, multidisciplinary optimization of an airfoil, and shape optimization of a dividing tube. By presenting sizing and shape optimization in an abstract way, the authors are able to use a unified approach in the mathematical analysis for a large class of optimization problems in various fields of physics. Sokolowski and Zolesio [SZ92] discussed the shape calculus introduced by J. Hadamard and extended it to a broad class of free boundary value problems. The approach is functional analytic throughout and will serve as a basis for the development of numerical algorithms to the solution of shape optimization problems. They dealt solely with sensitivity analysis but omitting approximation and computational aspects.

In the research area of shape optimization for piezoceramics, so far, no literature referred to in this dissertation concerns multi-objective optimization problems for piezecreamics.

Topology optimization. A large number of studies have been carried out in this subject. Topology optimization uses finite element methods to generate optimal design concepts. Topology optimization is comprised of five steps. First, the geometry of the design domain, boundary conditions and loading are prescribed. Second, the design domain is discretized by finite elements, and each element is assigned a design variable. Third, the finite element analysis and sensitivity analysis are used to give the function value and the first-order sensitivity of the objective and constraint. After that, the optimization algorithm is used to solve the optimization problem. Finally, the optimal topology is interpreted and refined.

Topology optimization with a homogenization method was proposed by Bensdøe and Kikuchi [BK88] to design a very stiff structure, and this method was then applied to design compliant mechanism and composite materials. Li et al. [LXKS01] developed a two layered optimization procedure (Topology Optimization and Genetic Algorithm Optimization) to solve a mixed optimization problem which designed both mechanical and electrical parts of a piezoelectric actuator. A topology optimization method is used to obtain the initial topology of the compliant mechanism, followed by detailed finite element analysis [BF04]. The effect of geometry parameters, material selection, and epoxy bonding layers in the piezoelectric actuator are also studied. Bürmann et al. [BRG03] optimized a piezoelectric fan with two symmetrically placed piezoelectric patches through an analytical Bernoulli-Euler model as well as a finite element (FE) model of the composite piezo-beam. Canfield and Frecker [CF00] designed the displacement amplifying compliant mechanisms for piezoelectric actuators by using a topology optimization approach. Two different solution methods, Sequential Linear Programming and an Optimality Criteria method, are used to optimize a two-objective problem. Allaire et al. [AJT02] studied a level-set method for numerical shape optimization of elastic structures. It combines the level-set algorithm with the classical shape gradient. This method can easily handle topology changes for a very large class of objective functions.

Although a great many studies have been reported, there is still more work worth to do, especially in the new research area of shape optimization for piezoceramics.

As a developing area, there are several interesting questions which need to be considered, e.g. how a shape of a piezo influences its (nonlinear) behavior? Does an unusual shape play an important role in improving its performance? Can these information be used in multi-objective optimization problems? With regard to these questions, the motivation of this work follows in Section 1.3.

1.3   Motivation of this Work

Considering the questions listed at the end of Section 1.2, the motivation of this work is proposed as follows: **to investigate the influence of the shape of a piezo on its (nonlinear) dynamical behavior and use this information for the shape optimization of the piezoceramics with respect to several objectives.**

Generally speaking, an optimization problem consists of two main major parts: a problem formulation, which includes the definition of design variables, objective function and constraints, and an optimization algorithm, which defines the numerical procedure with which the optimal solution is pursued. As mentioned in Section 1.2, shape optimization is still a new topic in the area of optimization problems for piezoelectric actuators. Since there is no established literature, shape optimization for piezoceramics is a challenging task for both mathematicians and engineers.

Project feasibility analysis To ensure an effective result, the feasibility of improving the response amplitude by changing the shape of a piezo with a computer software package Comsol Multiphsics (former FEMLAB, see Appendix A) was analyzed.

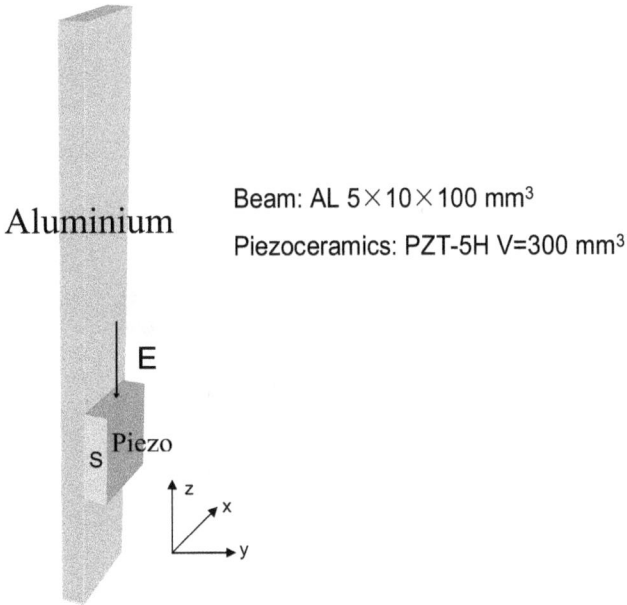

Beam: AL $5 \times 10 \times 100$ mm^3

Piezoceramics: PZT-5H V=300 mm^3

Figure 1.1: Example: bending of a beam

Figure 1.1 shows an example of a static analysis using piezoelectric volume elements. The use of a mode of piezoelectric materials has been investigated by Benjeddou et al. ([BTO97], [BTO99]).

The bender of a piezo-beam in Figure 1.1 is considered; it consists of an aluminium cantilever beam ($5 \times 10 \times 100\ mm^3$), which is fixed at the surface at $z = 0$, and a PZT-5H (Lead Zirconate Titanate) (Vol.$= 300\ mm^3$, contact area: $10 \times 10\ mm^2$) adjoined to the cantilever beam at a distance $15\ mm$ from the clamp. A 20 V potential difference is applied between the top and the bottom surfaces of the piezo. The material properties are summarized in Table 1.1.

Table 1.1: Material properties

Aluminium		
ρ 2690		(kg/m^3)
Y 70.3		(GPa)
ν 0.345		
PZT-5H		
ρ 7730		(kg/m^3)
$[c]$ $\begin{bmatrix} 126 & 79.5 & 84.1 & 0 & 0 & 0 \\ 79.5 & 126 & 84.1 & 0 & 0 & 0 \\ 84.1 & 84.1 & 126 & 0 & 0 & 0 \\ 0 & 0 & 0 & 23.3 & 0 & 0 \\ 0 & 0 & 0 & 0 & 23.0 & 0 \\ 0 & 0 & 0 & 0 & 0 & 23.0 \end{bmatrix}$		(GPa)
$[\varepsilon]$ $\begin{bmatrix} 1.503 & 0 & 0 \\ 0 & 1.503 & 0 \\ 0 & 0 & 1.3 \end{bmatrix} 10^{-8}$		(F/m)
$[e]$ $\begin{bmatrix} 0 & 0 & 0 & 0 & 0 & 17 \\ 0 & 0 & 0 & 0 & 17 & 0 \\ -6.5 & -6.5 & 23.3 & 0 & 0 & 0 \end{bmatrix}$		(Cb/m)

Figure 1.2 shows the application of Comsol Multiphsics for modeling piezoelectric effects in a static linear analysis using the **3D Electrostatics** application mode and the **3D**

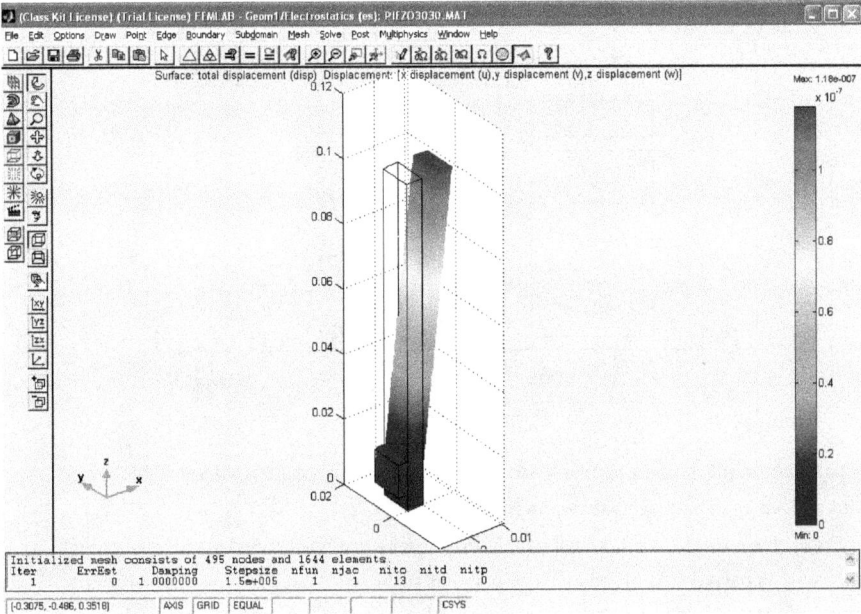

Figure 1.2: A static 3D example with Comsol Multiphsics (FEMLAB): bending of a beam (with a cuboid piezo)

Structural Mechanics Module Solid application mode. The mode of the piezoelectric material is used to accomplish a deflection of the tip. Table 1.2 shows the tip deflections for five different shapes of piezoceramics. The second column 'Shape of a piezo (S)' shows the shapes in two dimensionally as S in Figure 1.1.

In the first case, the piezo in Figure 1.2 is a cuboid ($3 \times 10 \times 10 \ mm^3$, $S = 3 \times 10 \ mm^2$). The corresponding tip deflection of the beam is $1.1814 \cdot 10^{-7} \ m$.

Then we change the shapes of the piezo to polygons. For some polygons, e.g. shapes No. 2 and 3 in Table 1.2, the results are decreased by around 1.5% when compared to the rectangular shape (No.1 in Table1.2). But in some cases, e.g. shape No.4, the tip deflection of the beam is $1.1952 \cdot 10^{-7} \ m$, an improvement of 1.1%.

9

Table 1.2: Beam's tip deflections for example shapes of piezoceramics

No.	Shape of a piezo (S)	Beam's tip deflection (10^{-7} m)
1		1.1814
2		1.1678
3		1.1601
4		1.1952
5		1.2085

As a third example, a shape similar to the No.4 but with a curved side (No.5 in Table 1.2) is introduced and the result is shown in Figure 1.3.

From the above results, we find that better performance can be obtained by changing the shape (e.g. a curved side instead of a rectangular shape) of a fixed-volume piezo-beam. Therefore, in this work we will focus on shapes like No.5 in Table 1.2.

According to the motivation above, the primary goals of this work are:

- finding a mathematical model to describe the problem and configure the design variables;

- deriving a general expression of the eigenfunctions for a piezo that can be used to describe the properties of piezoceramics for different shapes;

- choosing appropriate objective functions and constraints for the multi-objective shape optimization problem;

- solving the multi-objective optimization problem with an effective optimization method.

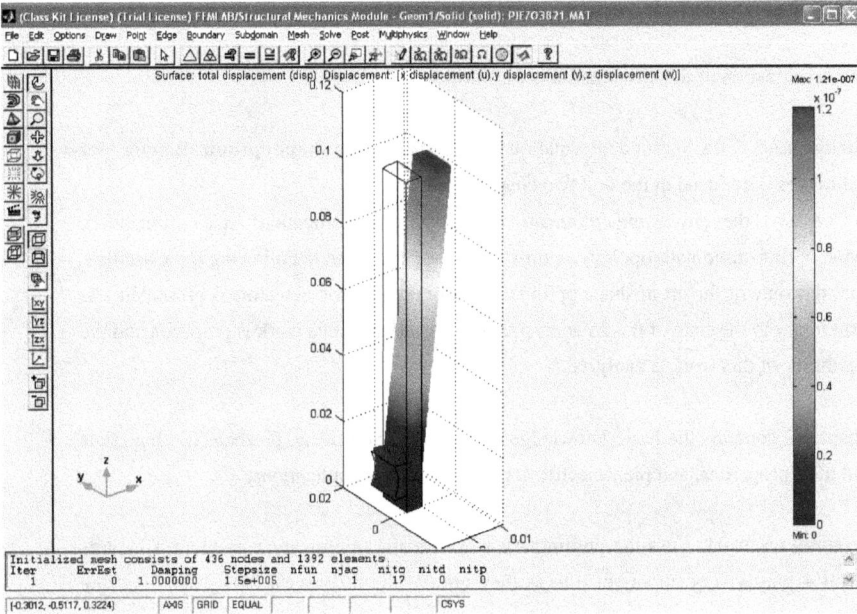

Figure 1.3: A static 3D example with Comsol Multiphsics (FEMLAB): bending of a beam (with a curved surface piezo)

In this dissertation, first we will introduce a mathematical model to describe a nonlinear phenomenon of piezoceramics observed in experiments; second, the general forms of the eigenfunctions for piezoceramics with different shapes will derived via a Hamilton's principle; then two objective functions and constraints are given; and at the end a numerical subdivision technique will be introduced to solve the multi-objective optimization problem. The compromised solutions for the multi-objective optimization problem are then obtained.

11

1.4   Organization of this Dissertation

The organization of this dissertation is as follows:

Background.   Background information for this dissertation **shape optimization** for **piezoecramics** is introduced in the first two chapters.

In Chapter 1, the term *shape optimization* and its three classifications (size optimization, shape optimization and topology optimization) are introduced. Following the classification, the state of the art in shape optimization of piezoelectric actuators is presented. As a summary to the state of the art analysis, the motivation of this work is proposed and the feasibility of this work is analyzed.

Chapter 2 contains the basic knowledge of piezoelectric effects, piezoelectric materials and their properties, and piezoelectric actuators and their applications.

Preliminary work.   As the preliminary work for the optimization problem, Chapters 3 and 4 play a very important role as they present the theoretical background and the essential computation for the further shape optimization problem.

In Chapter 3 we derived the general eigenfunctions of piezoceramics for different geometries of piezoelectric materials which describe how the shape (geometry) of a piezo influences its properties. Unusual shapes (compared to the rectangular shape) are considered in this chapter. A corresponding boundary value problem is solved numerically and the computation results are used to compute the coefficients of the equation of motion. Hamilton's principle is used to derive the linear and nonlinear equations of motion for a general shape. A Galerkin method is used to simplify the linear problem for a rectangular shape, and we assume this simplification can also be applied to a nonlinear model as well as a general shape (e.g a curved side shape).

In Chapter 4 equations of motion for both linear and nonlinear cases are solved numerically via a continuation software package AUTO2000. Results show that better performance can be obtained with certain unusual shapes. A nonlinear dynamical behavior of piezoceramics under weak electric fields is introduced and the influences of the nonlinear terms are analyzed.

12

Optimization.   A multi-objective shape optimization problem for piezoceramics is introduced in Chapter 5, and optimization results are discussed in Chapter 6.

In Chapter 5 multi-objective optimization problems are introduced and the main optimization techniques currently available are reviewed. Particularly, a set oriented multilevel subdivision technique is studied. The formulation of a multi-objective shape optimization problem and a GAIO (Global Analysis of Invariant Objects) model for solving the optimization problem are given.

In Chapter 6 the multi-objective shape optimization results are discussed. Two cases, a quadratic curve (one design variable) and a cubic B-spline curve (two design variables), are considered. For both cases Pareto sets are obtained. Results show that the performance of piezoceramics can be improved by changing their shapes.

Conclusion   In Chapter 7 a summary of this dissertation and an outlook for the possible research work in the future are given.

1 Introduction

Chapter 2

Piezoelectricity

In this chapter, a short history of piezoelectricity, the direct and inverse piezoelectric effects, and piezoelectric constitutive equations are presented in Section 2.1. In Section 2.2 piezoelectric materials and their properties are introduced. The piezoelectric actuators and their applications are briefly introduced in Section 2.3.

2.1 Piezoelectric Effect

2.1.1 Introduction

The piezoelectric effect was first mentioned in 1817 by the French mineralogist René Just Haüy. It was first demonstrated by Pierre and Jacques Curie in 1880. They found that if certain crystals were subjected to mechanical strain, they became electrically polarized and the degree of polarization was proportional to the applied strain. The Curies also discovered that these same materials deformed when they were exposed to an electric field. This has become known as the inverse piezoelectric effect [Pie01].

Their experiments led them to elaborate on the early theory of piezoelectricity. This theory was complemented by the further work of Lippman, Hankel, Kelvin and Voigt (beginning of 20th century). Hankel proposed the name 'piezoelectricity[1]'. Until the beginning of the century, the piezoelectricity did not leave laboratories. The first practical use of the piezoelectric effect was during the first World War when sonar emitters (P. Langevin) were

---

[1]The prefix 'piezo-' is derived from the Greek word piezein, meaning to press or squeeze.

effectively used to detect German submarines by producing ultrasonic waves with piezo-electric quartz. Prior to the second World War, researchers at MIT discovered that certain ceramics such as PZT (lead zirconate titanate) could be polarized to yield a high piezo response. In the twenties, the use of quartz to control the resonance frequency of oscillators was proposed by an American physicist W. G. Cady. It is during the period following the first world war that most of the piezoelectric applications we are now familiar with (microphones, accelerometers, ultrasonic transducers, benders, etc.) were conceived. However, the materials available at the time often limited device performance. The development of electronics, especially during the second World War, and the discovery of ferroelectric ceramics increased the use of piezoelectric materials. The direct piezoelectric effect consists of the ability of certain crystalline materials (i.e. ceramics) to generate an electrical charge in proportion to an externally applied force. The direct piezoelectric effect has been widely used in transducer design (accelerometers, force and pressure transducers, etc.). According to the inverse piezoelectric effect, an electric field induces a deformation of the piezoelectric material. The inverse piezoelectric effect has been applied in actuator design [Ike90]. Figure 2.1 is the schematic diagram of piezoelectric effects.

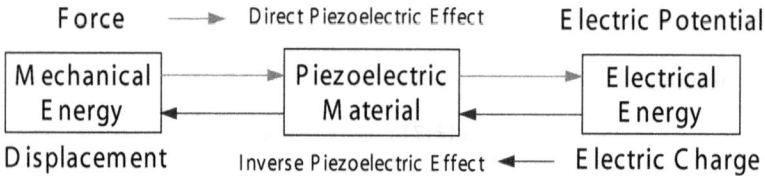

Figure 2.1: schematic diagram of piezoelectric effects

In a piezoelectric crystal, the positive and negative electrical charges are separated, but symmetrically distributed, so that the crystal is overall electrically neutral. When stress is applied, this symmetry is destroyed, and the asymmetry charge generates a voltage. The converse piezoelectricity is where application of an electrical field creates mechanical stress (distortion) in the crystal. Because the charges inside the crystal are separated, the applied voltage affects different points within the crystal differently, resulting in distortion. In the simplest of terms, when a piezoelectric material is squeezed, an electric charge collects on its surface. Conversely, when a piezoelectric material is subjected to a voltage drop, it mechanically deforms.

2.1.2   Piezoelectric Constitutive Equations

What is a constitutive equation? For mechanical problems, a constitutive equation describes how a material strains when it is stressed, or vice-versa. Constitutive equations also exist for electrical problems; they describe how charges move in a (dielectric) material when it is subjected to voltage, or vice-versa.

Engineers are already familiar with the most common mechanical constitutive equation that applies for usual metals and plastics. This equation is known as Hooke's Law and is written as:

$$S = s \cdot T$$

In words, this equation states: Strain = Compliance × Stress.

However, since piezoelectric materials are concerned with electrical properties too, we must also consider the constitutive equation for common dielectrics:

$$D = \epsilon \cdot E$$

In words, this equation states: Charge Density = Permittivity × Electric Field.

Piezoelectric materials combine these two seemingly dissimilar constitutive equations into one coupled equation, which defines how the piezoelectric material's stress (T), strain (S), charge-density displacement (D), and electric field (E) interact.

The piezoelectric constitutive law (in Strain-Charge form) is:

$$S = s_E T + d^t E \tag{2.1}$$

$$D = d\,T + \epsilon_T E \tag{2.2}$$

The matrix d contains the piezoelectric coefficients for the material, and it appears twice in the constitutive equation (the superscript t stands for matrix-transpose). The subscripts in piezoelectric constitutive equations have very important meanings. They describe the conditions under which the material property data was measured. For example, the subscript E on the compliance matrix s_E means that the compliance data was measured under at least a constant, and preferably a zero, electric field. Likewise, the subscript T on the permittivity matrix ϵ_T means that the permittivity data was measured under at least a constant, and preferably a zero, stress field.

The four state variables (S, T, D, and E) can be rearranged to give 3 additional forms for a piezoelectric constitutive equation. Instead of the coupling matrix d, they contain the coupling matrices e, g, or q. It is possible to transform piezo constitutive data from one form to another.

2.2 Piezoelectric Materials

The piezoelectric effects can be seen as transfers between electrical and mechanical energy. Such transfers can only occur if the material is composed of charged particles and can be polarized. For a material to exhibit an anisotropic property such as piezoelectricity, its crystal structure must have no center of symmetry. 21 crystal structures out of 32 are non-centrosymmetric. A crystal having no center of symmetry possesses one or more crystallographically unique directional axes. All 21 non-centrosymmetric crystal classes, except one, show piezoelectric effect along the directional axes. Out of the 20 piezoelectric classes, 10 have only one unique direction axis. Such crystals are called polar crystals as they show spontaneous polarization. The value of the spontaneous polarization depends on the temperature. This is called the pyroelectric effect. The pyroelectric crystals for which the magnitude and direction of the spontaneous polarization can be reversed by an external electric field are said to show ferroelectric behavior. Most of the piezoelectric materials are crystalline solids. They can be single crystals, either formed naturally or by synthetic processes, or polycrystalline materials like ferroelectric ceramics which can be rendered piezoelectric and given, on a macroscopic scale, a single crystal symmetry by the process of poling (by subjecting to a high electric field not far below the Curie temperature). The piezoelectric effect can also appear in crystals composed of only one type of element (in this case, the polarization is due to a distortion in the electronic distribution). Certain polymers can also be made by stretching them under an electrical field.

Material properties. It is well known that the mechanical and electrical responses of a piezoelectric material are coupled. When the applied electric field is low and the strains are also low, the behavior of piezoceramics is almost linear. However, a wide range of nonlinear piezoelectric phenomena are observed both under high and low electric fields. The nonlinear behaviors under high and low electric fields are different in some aspects. For example, the nonlinearities observed under high electric fields are hysteresis behavior between the electric field and strain, nonlinear relation between electric field and mechanical displacement, etc. The nonlinearities observed under weak electric fields are jump phenomena, dependance of resonance frequency on vibration amplitude, presence of superharmonics in the response spectra and nonlinear relationship between applied electric voltage and mechanical displacement, etc.[Sea05].

Polarization is the amount of charge associated with the dipolar or free charge in a dielectric. Figure 2.2 shows schematically the domain reorientation in a multi-domain piezoelectric material.

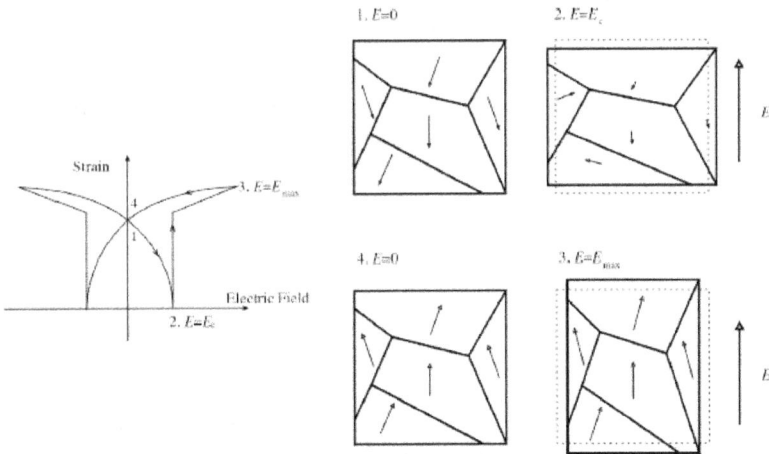

Figure 2.2: Strain change associated with the polarization reorientation (adapted from [Pie01])

19

The material is initially poled along the negative direction (1). When an electric field is applied along the positive direction, the crystal will first shrink with the increase of the field since the field is opposite in direction to the polarization. The strain reaches a minimum at a certain field (coercive field E_c), where the polarization starts to reverse in each grain (2). Above E_c, the crystal expands until E_{max} as the field now has the same direction as the polarization. Near E_{max}, all the reversible polarization has been reversed. As the field is reduced, the strain decreases monotonically as no polarization reversal occurs. The situation for a zero electric field (4) is similar to the starting situation except that the polarization is reversed; the material is now poled along the positive direction.

Since the piezo effect exhibited by natural materials such as quartz, tourmaline, Rochelle salt, etc. is very small, polycrystalline ferroelectric ceramic materials such as BaTiO3 and Lead Zirconate Titanate (Piezo) have been developed with improved properties. Ferroelectric ceramics become piezoelectric when poled. Piezoceramics are available in many variations and are still the most widely used materials for actuator or sensor applications today. Piezo crystallites are centro-symmetric cubic (isotropic) before poling and after poling exhibit tetragonal symmetry (anisotropic structure) below the Curie temperature (see Figure 2.3). Above this temperature they lose their piezoelectric properties.

The vast majority of piezoelectric materials found in the marketplace today are inorganic ceramics such as lead titanate (PbTiO3), lead zirconium titanate (PbZrTiO3), lithium tantalate (LiTaO3), and barium titanate (BaTiO3). Most of these perovskite materials were pioneered in the late 1940's and 1950's, and are characterized by high elastic moduli, high dielectric constant, low elastic and dielectric loss, and high electro-mechanical coupling factors [Tho02].

Although modern piezoelectric ceramic materials have proven successful in many applications, they have a number of inherent limitations:

1. Low yield strains eliminate ceramics from high strain sensing applications such as flexure in helicopter rotor blades and in fishing rods. Biomedical applications such as foot strike force or bite pressure are also impractical for ceramic materials.

2. Brittleness makes these materials prone to fracture and crack propagation, and makes the use of ceramic materials in impact, shock, and ordinance applications

(1) (2)

Pb O²⁻ Ti, Zr

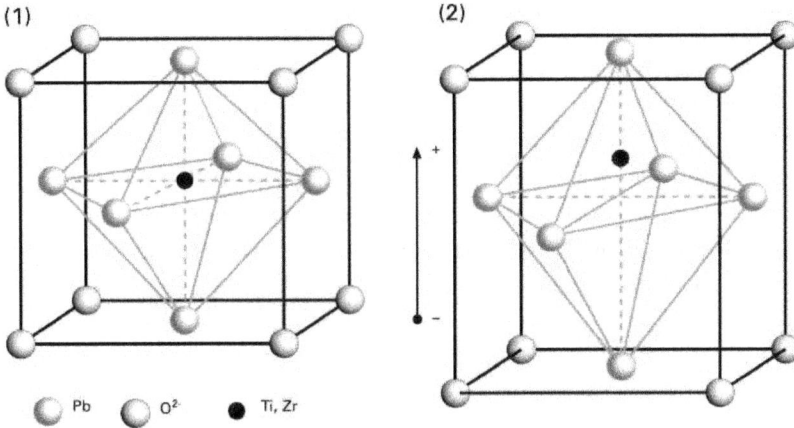

Figure 2.3: Piezoelectric elementary cell (a) before poling (b) after poling

impractical. Brittle piezoelectric materials used in situations undergoing positive and negative stresses must often be preloaded into compression to ensure mechanical stability.

3. The density of ceramic materials is high, creating problems for weight sensitive applications such as naval hull mounted and geophysical towed array sonar systems.

4. The acoustic impedance of ceramic materials (a function of density and stiffness) is high, providing poor acoustic coupling to lower impedance materials like water or human tissue.

5. The costs of the raw materials, basic processing of piezoelectric ceramics and single crystal materials are relatively high per unit volume.

2.3 Piezoelectric Actuator

A Piezoelectric actuator is a device that uses ceramics with piezoelectric characteristics to produce movement. It converts electrical input energy into an output such as a displacement or generated force.

Piezoelectric actuators have been widely used in various fields such as micro-positioning of tools, active vibration control, ultrasonic welding and machining, and common rail diesel injection systems. Table 2.1 shows some applications of piezoelectric actuators. They are used in high speed, high-accuracy control of valves in semiconductor manufacturing equipment, in ultra-precise positioning and in the generation and handling of high forces or pressures in static or dynamic situations. They can also be used in optical switches that move tiny mirrors and in endoscopic lenses used in medical treatment. To satisfy stricter governmental regulations with concerns to air pollution, one promising application area is for motor vehicles to reduce emission of particulate matter to fulfill the environment regulations. The piezoelectric effect is used in sensing applications, such as in force or displacement sensors. The inverse piezoelectric effect is used in actuation applications, such as in motors and devices that precisely control positioning, and in generating sonic and ultrasonic signals. For most piezoelectric actuators, as for sensors, a reasonably linear relationship between input signal and movement is required. However, there is the special class of actuators, which is purposely driven at their resonant frequency, known as ultrasonic transducers.

Table 2.1: Applications of piezoelectric actuators

areas	application	notes
Medical devices	surgical tools and ultrasonic testing	
Micropositioning & Nanopositioning	specific application requirements including tight geometries, weight restrictions, and vacuum and non-magnetic constructions.	1. the fastest responding positioning element available with microsecond time constants; 2. producing motions in sub-nanometer increments.
Spacecraft instrumentation	precise positioning of optics; active damping of vibration	
Piezoelectric motors [Mor03]	linear motors, rotational motors	
Vibration control of civil structures	the application of piezoceramic actuators in various civil structures such as beams, trusses, steel frames and cable-stayed bridges	1. applications related to civil engineering; 2. low-cost, lightweight, and easy-to-implement materials for active control of structural vibration.

23

2   Piezoelectricity

Chapter 3

Derivation of General Eigenfunctions and Equations of Motion

When the shape of a piezo is changed, the eigenfunctions and the eigenfrequency are changed correspondingly. Therefore the tasks of this chapter arise: to derive the general eigenfunctions of piezoceramics with different shapes, and to solve the corresponding boundary value problem and use the computation results for a further optimization problem. In the first section, an unusual example shape is considered with one boundary described by a curved side $y(z)$ instead of a straight line (special case of $y(z)$ is a constant). Then a nonlinear model is introduced to describe the nonlinear behavior of piezoceramics excited by weak electric fields. In the second section, the general eigenfunctions of piezoceramics with different geometries are derived for a linear model and the corresponding boundary value problem is solved numerically. Then an ansatz is proposed to compute the spatial eigenfunctions of a piezo. The linear and nonlinear equations of motion are derived for the general shape in Section 3.3. At the end of this chapter, a Galerkin method is used to simplify the ansatz for a rectangular shape with a linear model, and we assume this simplification can also be applied for a nonlinear model as well as a general shape (e.g a curved side shape).

3.1   A Nonlinear Model

In this section, the example shape of piezoelectric material is given and the nonlinear corresponding model is introduced.

The example shape.   As a concrete example we consider a piezo as depicted in Figure 3.1 [DJW05]. Between the top ($z = l/2$) and the bottom ($z = -l/2$) an alternating external voltage with amplitude U_0 is applied. The piezo vibrates with an amplitude in z direction. In structural theory, the members of a structure are not, in general, treated as three-dimensional continua but rather as continua of one or two dimensions. To simplify the problem we consider it in two dimensions and only part of the boundary of the piezo is subject to change [HM03]. This part is parameterized by $y(z)$.

Definition 3.1.1. *Let the shape* $\mathcal{S} \subseteq \mathbf{R}^2$ *of a piezo be given by*

$$\mathcal{S} = \{(\bar{y}, \bar{z}) | \frac{-l}{2} \leq \bar{z} \leq \frac{l}{2}, 0 < \bar{y} \leq y(\bar{z})\}$$

with a function $y : \mathbf{R} \to \mathbf{R}$*. Then* y *is called the shape function of a piezo.*

Remark 3.1.2.   The piezo shape function $y(z)$ depicted in Figure 3.1 is parameterized in a very simple way. In real-life problems, however, the parametrization of the geometry may be a difficult task.

A nonlinear model.   There is a wide range of nonlinear effects which can be observed in piezoceramics. One example is the well-known butterfly behavior for large stresses and electric fields. For small stresses and weak electric fields, piezoceramics are usually described by linear constitutive equations around an operating point in the butterfly hysteresis curve. Nevertheless, typical nonlinear effects can be observed when piezoelectric actuators and structures with embedded piezoceramics are excited in resonance even if the electric field remains small. This was observed and described by Beige and Schmidt in 1982. They modeled these nonlinearities using higher order quadratic and cubic elastic and electric terms. Typical nonlinear effects, e.g. a nonlinear relation between excitation voltage and vibration amplitude, were also observed. Based on the observed experimental behavior, von Wagner et al. developed theoretical models for the electric enthalpy

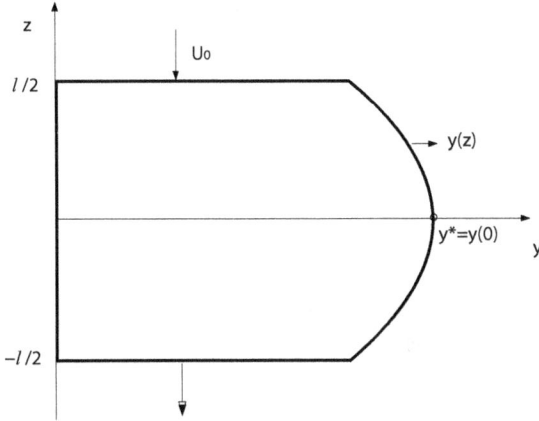

Figure 3.1: Shape of a piezo under consideration.

density function, which is subsequently used in Hamilton's principle to derive governing equations for the piezo-continuum [vWH03].

To derive equations of motion for a nonlinear model, we use Lagrange-d'Alembert principle

$$\delta \int_{t_0}^{t_1} \mathrm{L}\,dt + \int_{t_0}^{t_1} \delta \mathrm{W}\,dt = 0. \tag{3.1}$$

where t_0 and t_1 define the time interval (all variations must vanish at $t = t_0$ and $t = t_1$), L is the Lagrangian

$$\mathrm{L} \;=\; \int_v (\mathrm{T} - \mathrm{H})\,dv$$

$$=\; \int_z A\big(\mathrm{T}(\dot{w}(z,t)) - \mathrm{H}(w'(z,t), \varphi'(z,t))\big)\,dz,$$

where

$$(\,)' = \frac{\partial}{\partial z}, \quad (\dot{\,}) = \frac{\partial}{\partial t}.$$

27

$w(z,t)$ and $\varphi(z,t)$ are vertical displacement and electric potential respectively. T and H respectively denote the kinetic energy density and the electric enthalpy density. dv is the volume variation. A is the (constant) cross section of the piezo.

δW is virtual work done by external mechanical and electrical forces. Considering Neumann's work [Neu02], in this dissertation we introduce δW as

$$\delta\mathrm{W} = -\int_{-l/2}^{l/2} A\left(E_d^0 \dot{w}' \delta w' + E_d^2 \dot{w}'^3 \delta w'\right) dz, \qquad (3.2)$$

where E_d^0 and E_d^2 are obtained experimentally.

In [vWH03], the kinetic energy is expressed by

$$T(\dot{w}) = \frac{1}{2}\rho \dot{w}^2(z,t),$$

where ρ is the density.

The electric enthalpy density including higher order terms is also given in von Wagner's work [vWH03]:

$$
\begin{aligned}
H = \;& \frac{1}{2}E^0 S_{zz}^2 - \gamma_0 S_{zz} E_z - \frac{1}{2}\nu_0 E_z^2 \\
&+\frac{1}{4}E^2 S_{zz}^4 - \frac{1}{3}\gamma_2^1 S_{zz}^3 E_z - \frac{1}{2}\gamma_2^2 S_{zz}^2 E_z^2 - \frac{1}{3}\gamma_2^3 S_{zz} E_z^3 - \frac{1}{4}\nu_2 E_z^4 \\
&+\frac{1}{3}E^1 S_{zz}^3 - \frac{1}{2}S_{zz}^2 E_z - \frac{1}{2}\gamma_1^2 S_{zz} E_z^2 - \frac{1}{3}\nu_1 E_z^3
\end{aligned}
\qquad (3.3)
$$

with

$$\nu_0 = \varepsilon_{33}^T - d_{33}^2 E^0, \quad \gamma_0 = E^0 d_{33}$$

where E^0 is Young's modulus. S_{zz} is the strain and E_z is the electric field in the z-direction respectively. The parameter d_{33} corresponds to the piezoelectric 33-effect and ε_{33}^T is the dielectric constant measured at constant stress. The terms containing E^0, γ_0 and ν_0 correspond to the classic linear theory. The fourth order terms containing E^2, $\gamma_2^1, \gamma_2^2, \gamma_2^3$ and ν_2 will produce cubic nonlinearities in the equations of motion and the third order terms containing E^1, γ_1^1, γ_1^2 and ν_1 will produce quadratic nonlinearities. The term including ν_2 is due to the effect of electrostriction [vWH03].

Using the kinematic relation

$$S_{zz}(z,t) = w'(z,t),$$

and the electric potential φ given by

$$E_z = -\varphi'(z,t),$$

we obtained the following equation for a piezoelectric rod via equation (3.1):

$$\delta \int_{t_0}^{t_1} \int_{-l/2}^{l/2} A\{\frac{1}{2}\rho\dot{w}^2 - \frac{1}{2}E^0 w'^2 - \gamma_0 w'\varphi' + \frac{1}{2}\nu_0\varphi'^2$$

$$-\frac{1}{4}E^2 w'^4 - \frac{1}{3}\gamma_2^1 w'^3\varphi' + \frac{1}{2}\gamma_2^2 w'^2\varphi'^2 - \frac{1}{3}\gamma_2^3 w'\varphi'^3 + \frac{1}{4}\nu_2\varphi'^4$$

$$-\frac{1}{3}E^1 w'^3 - \frac{1}{2}\gamma_1^1 w'^2\varphi' + \frac{1}{2}\gamma_1^2 w'\varphi'^2 - \frac{1}{3}\nu_1\varphi'^3\} \, dzdt$$

$$-\int_{t_0}^{t_1} \int_{-l/2}^{l/2} A\{E_d^0 \dot{w}'\delta w' + E_d^2 \dot{w}'^3\delta w'\} \, dzdt \quad = \quad 0. \qquad (3.4)$$

The external voltage applied at the top and the bottom of the piezo is predetermined, i.e.

$$\varphi(l/2, t) - \varphi(-l/2, t) = U_0 \cos\Omega t$$

holds at any time. If the potential of the electrode at $z = -l/2$ is set as $-\frac{U_0}{2}\cos\Omega t$, the potential of the electrode at $z = l/2$ holds as $\frac{U_0}{2}\cos\Omega t$. This leads to two boundary conditions:

$$\varphi(l/2, t) \quad = \quad \frac{U_0}{2}\cos\Omega t, \qquad (3.5)$$

$$\varphi(-l/2, t) \quad = \quad -\frac{U_0}{2}\cos\Omega t. \qquad (3.6)$$

These conditions are valid for both the linear and the nonlinear models.

3.2 General Eigenfunctions Derivation for Piezoceramics with a Linear Model

In this section, the eigenfunctions of the linear problem with short circuited electrodes ($U_0 = 0$) are calculated in order to obtain suitable shape functions. In Section 3.2.1 the linear model for different geometries of piezoceramics is obtained by setting all the

nonlinear parameters in equation (3.4) equal to zero. In Section 3.2.2, the general eigenfunctions are derived and the results are obtained by solving a boundary value problem
numerically. Finally , the computations for different $y(z)$ are given in Section 3.2.3.

3.2.1   A Linear Model

Extending the work of von Wagner, we consider the case that A (the cross section of a
piezo) in equation (3.4) is not constant. Concretely, we allow y in Figure 3.1 to explicitly depend on z. When $y(z)$ is no longer a constant, the shape of the piezo changes
correspondingly, as do the eigenfunctions and the eigenfrequency.

A linear model is obtained when we consider that virtual work $\delta \mathrm{W}$ vanishes in the un-
damped case

$$\delta \mathrm{W} = 0,$$

and set the nonlinear parameters in equation (3.4) equal to zero. Then equation (3.4) is
simplified to

$$\delta \int_{t_0}^{t_1} \int_{-l/2}^{l/2} y(z) \underbrace{\left(\frac{1}{2}\rho\dot{w}^2 - \frac{1}{2}E^0 w'^2 - \gamma_0 w'\varphi' + \frac{1}{2}\nu_0\varphi'^2 \right)}_{F(w',\varphi',\dot{w})} dzdt = 0. \qquad (3.7)$$

Proposition 3.2.1. *For the linear equation 3.7, if a piezo is described by a piezo shape
function* $y(z) : \mathbf{R} \to \mathbf{R}$*, then the vertical displacement* $w(z,t)$ *satisfies the partial differential equation*

$$E^*(y(z)w'' + y'(z)w') = \rho y(z)\ddot{w}, \qquad (3.8)$$

and the relationship between $w(z,t)$ *and the electric potential* $\varphi(z,t)$ *is expressed through
the fact that* $\varphi(z,t)$ *satisfies the equation*

$$y(z)\varphi'' + y'(z)\varphi' = \alpha(y(z)w'' + y'(z)w'), \qquad (3.9)$$

where

$$\alpha = \frac{\gamma_0}{\nu_0}, \quad E^* = E^0 + \frac{\gamma_0^2}{\nu_0}.$$

Proof. According to the Calculus of Variations [Arn78], we know that J in equation 3.7 has an extremum only if the Euler-Lagrange differential equation is satisfied, e.g.

$$\begin{cases} F_w - \frac{\partial}{\partial z}F_{w'} - \frac{\partial}{\partial t}F_{\dot{w}} & = \quad 0 \\ F_{\varphi} - \frac{\partial}{\partial z}F_{\varphi'} - \frac{\partial}{\partial t}F_{\dot{\varphi}} & = \quad 0. \end{cases} \tag{3.10}$$

Now we consider the equations in (3.10) respectively. Since $F_{\varphi} = 0$ and $F_{\dot{\varphi}} = 0$, for the second equation in (3.10) we obtain

$$-\frac{\partial}{\partial z}F_{\varphi'} \qquad\qquad = 0$$
$$\Rightarrow \qquad -\frac{\partial}{\partial z}\left(y(z)(-\gamma_0 w' + \nu_0 \varphi')\right) \qquad = 0$$
$$\Rightarrow \quad y'(z)(\gamma_0 w' - \nu_0 \varphi') + y(z)(\gamma_0 w'' - \nu_0 \varphi'') \quad = 0. \tag{3.11}$$

equation (3.11) can be simplified to

$$y(z)\varphi'' + y'(z)\varphi' = \alpha(y(z)w'' + y'(z)w').$$

To prove the equation (3.8) in Proposition 3.2.1, we consider the first equation in (3.10), for which we have $F_w = 0$. Then the equation can be rewritten to

$$-\frac{\partial}{\partial z}F_{w'} - \frac{\partial}{\partial t}F_{\dot{w}} \qquad\qquad = 0$$
$$\Rightarrow \quad -\frac{\partial}{\partial z}\left(y(z)(-E^0 w' - \gamma_0 \varphi') - \frac{\partial}{\partial t}\rho \dot{w}y(z)\right) \quad = 0$$
$$\Rightarrow \quad y'(z)(E^0 w' + \gamma_0 \varphi') + y(z)(E^0 w'' + \gamma_0 \varphi'') \quad = \rho \ddot{w}y(z). \tag{3.12}$$

Substituting equation (3.9) into (3.12), we obtain the statement:

$$E^*(y(z)w'' + y'(z)w') = \rho y(z)\ddot{w}.$$

$\square$

Remark 3.2.2. In Proposition 3.2.1,

1. $w(z, t)$ is a solution of the partial differential equation (3.8) which we will solve using eigenfunctions of the corresponding differential operator.

2. In Section 3.2.2 we will use (3.9) to compute the corresponding solutions for $\varphi(z, t)$ since (3.9) shows a relation between $\varphi(z, t)$ and $w(z, t)$.

To derive boundary conditions necessary for the solution of equation (3.8), we follow the work of von Wagner [vWH02] and expand the variational equation (3.7) in terms of $\delta\varphi$ and δw.

Proposition 3.2.3. *In the situation of Proposition 3.2.1, the vertical displacement $w(z,t)$ and the electric potential $\varphi(z,t)$ satisfy the boundary conditions*

$$E^0 w'(l/2, t) + \gamma_0 \varphi'(l/2, t) = 0 \qquad (3.13)$$

$$E^0 w'(-l/2, t) + \gamma_0 \varphi'(-l/2, t) = 0. \qquad (3.14)$$

Proof. The equation (3.7) can be rewritten to

$$\int_{t_0}^{t_1} \int_{-l/2}^{l/2} \delta F(w', \dot{w}, \varphi') dz dt = 0$$

$$\Rightarrow \int_{t_0}^{t_1} \int_{-l/2}^{l/2} \left(F_{w'} \delta w' + F_{\varphi'} \delta \varphi' \right) dz dt + \int_{-l/2}^{l/2} \int_{t_0}^{t_1} F_{\dot{w}} \delta \dot{w} dt dz = 0 \qquad (3.15)$$

Performing integration by parts on equation (3.15) with respect to z and t, we obtain

$$\int_{t_0}^{t_1} \left[F_{w'} \delta w + F_{\varphi'} \delta \varphi \right]_{-l/2}^{l/2} dt - \int_{t_0}^{t_1} \int_{-l/2}^{l/2} \left(\frac{\partial}{\partial z} F_{w'} \delta w + \frac{\partial}{\partial z} F_{\varphi'} \delta \varphi \right) dz dt$$

$$+ \int_{-l/2}^{l/2} \left[F_{\dot{w}} \delta w \right]_{t_0}^{t_1} dz - \int_{-l/2}^{l/2} \int_{t_0}^{t_1} \left(\frac{\partial}{\partial t} F_{\dot{w}} \delta w \right) dt dz = 0$$

and rearrange it as

$$\int_{t_0}^{t_1} \left[F_{w'} \delta w + F_{\varphi'} \delta \varphi \right]_{-l/2}^{l/2} dt + \int_{-l/2}^{l/2} \left[F_{\dot{w}} \delta w \right]_{t_0}^{t_1} dz$$

$$- \int_{t_0}^{t_1} \int_{-l/2}^{l/2} \left(\frac{\partial}{\partial z} F_{w'} + \frac{\partial}{\partial t} F_{\dot{w}} \right) \delta w dz dt - \int_{t_0}^{t_1} \int_{-l/2}^{l/2} \frac{\partial}{\partial z} F_{\varphi'} \delta \varphi dz dt = 0. \qquad (3.16)$$

In equation (3.16), the third and fourth parts are the same as in equation (3.10) and have to equal zero. According to Hamilton's principle, $\delta w(z, t_0) = \delta w(z, t_1) = 0$ hold, then the second part is also zero. So equation (3.16) can be simplified to:

$$\int_{t_0}^{t_1} \left[\delta w F_{w'} + \delta \varphi F_{\varphi'} \right]_{-l/2}^{l/2} dt = 0 \tag{3.17}$$

For arbitrary t_0 and t_1, the time-integral in equation (3.17) has to vanish. Therefore the integrand has to vanish.

Since the electric potential at the electrodes of the piezo is predetermined (reminding of the two boundary conditions (3.5) and (3.6)), the variation of the electric potential with respect to z at the electrodes has to vanish, so

$$\delta \varphi(\pm \frac{l}{2}, t) = 0 \tag{3.18}$$

holds.

As the piezo is suspended freely at both ends, the displacements at $z = \pm l/2$ are not predetermined. Therefore $\delta w(\pm \frac{l}{2}, t)$ are not always zero.

Relating that $\delta \varphi(\pm \frac{l}{2}, t)$ equal zero and $\delta w(\pm \frac{l}{2}, t)$ are not always zero to equation (3.17), we obtain

$$F_{w'}(l/2, t) = F_{w'}(-l/2, t) = 0. \tag{3.19}$$

This leads to the two boundary conditions :

$$
\begin{aligned}
E^0 w'(l/2, t) + \gamma_0 \varphi'(l/2, t) &= 0 \\
E^0 w'(-l/2, t) + \gamma_0 \varphi'(-l/2, t) &= 0.
\end{aligned}
$$

$\square$

3.2.2 General Eigenfunctions Derivation for Different Geometries of Piezoceramics

In this section, we will derive the general eigenfunctions $W_k(z)$ and the corresponding functions $\Phi_k(z)$ that will be used to describe the electric potential φ. To do this, we first derive the relation between solutions $w(z, t)$ of (3.8) and the corresponding $\varphi(z, t)$ via equation (3.9).

Proposition 3.2.4. *Let the piezo shape function $y(z)$ be an even function, then*

1. *$w(z,t)$ is an odd function; and*

2. *one has*

$$\varphi(z,t) = \alpha(w(z,t) - w(l/2,t)f(z)) + \frac{U_0}{2}f(z)\cos\Omega t, \qquad (3.20)$$

   where

$$f(z) = \frac{g(z)}{g(\frac{l}{2})}$$

   and

$$g(z) = \int_0^z \frac{1}{y(s)}ds.$$

Proof.   1. Performing integration by parts on equation (3.9) with respect to z, we obtain

$$y(z)\varphi''(z,t) + y'(z)\varphi'(z,t) = \alpha(y(z)w'(z,t) + y'(z)w'(z,t))$$
$$\Rightarrow \qquad (y(z)\varphi'(z,t))' = \alpha\,(y(z)w'(z,t))'$$
$$\Rightarrow \qquad y(z)\varphi'(z,t) = \alpha y(z)w'(z,t) + D_1(t)$$
$$\Rightarrow \qquad \varphi'(z,t) = \alpha w'(z,t) + \frac{D_1(t)}{y(z)}$$
$$\Rightarrow \qquad \varphi(z,t) = \alpha w(z,t) + D_1(t)g(z) + D_2(t), \quad (3.21)$$

where $D_1(t)$ and $D_2(t)$ are functions of t.

Now we prove that $g(z)$ is odd. Since $y(z)$ is even as we defined in section 3.1, $y(z) = y(-z)$. We have

$$g(-z) = \int_0^{-z} \frac{1}{y(s)}ds = \int_0^z \frac{1}{y(-s)}d(-s) = -\int_0^z \frac{1}{y(-s)}ds$$
$$= -g(z).$$

So $g(z)$ is an odd function, and that $f(z)$ (see Proposition 3.2.4) is odd can be proved correspondingly.

Substituting equation (3.21) into the boundary conditions (3.13) and (3.14), we obtain

$$E^0 w'(l/2, t) + \gamma_0(w'(l/2, t) + D_1(t)g'(l/2)) \quad = 0$$
$$E^0 w'(-l/2, t) + \gamma_0(w'(-l/2, t) + D_1(t)g'(-l/2)) \quad = 0$$

(3.22)

We have already proven that $g(z)$ is an odd function, so $g'(z)$ is even. From equation (3.22) we additionally have:

$$w'(l/2, t) = w'(-l/2, t).$$

(3.23)

It can be proven that for a solution $w(z)$ to the equation (3.8), also $w(-z)$ is a solution. If the eigenvalue corresponding with $w(z)$ is simple, one therefore has

$$w(z) = \beta w(-z),$$

(3.24)

where β is a constant.

Observing that

$$
\begin{aligned}
w(z) &= \beta w(-z) = \beta^2 w\left(-(-z)\right) \\
&= \beta^2 w(z),
\end{aligned}
$$

we know that $\beta^2 = 1$ and therefore

$$\beta = \pm 1.$$

So $w(z)$ is either an even or odd function.

Taking into account that by equation (3.23) $w'(z, t)$ is even, $w(z, t)$ is odd. We will need this result to prove that $D_2(t)$ in equation (3.21) is zero.

2. Substituting the boundary conditions (3.5) and (3.6) into equation (3.21), we have:

$$\alpha w(l/2, t) + D_1(t)g(l/2) + D_2(t) \quad = \tfrac{U_0}{2}\cos \Omega t,$$
$$\alpha w(-l/2, t) + D_1(t)g(-l/2) + D_2(t) \quad = -\tfrac{U_0}{2}\cos \Omega t.$$

(3.25)

Adding these two equations in (3.25), one sees that $D_2(t)$ in (3.25) is zero since $g(z)$ and $w(z,t)$ are odd. Therefore equation (3.21) can be simplified to

$$\varphi(z,t) = \alpha w(z,t) + D_1(t)g(z). \tag{3.26}$$

Substituting equation (3.26) into the boundary condition (3.5)

$$\alpha w(l/2,t) + D_1(t)g(l/2) \;=\; \frac{U_0}{2}\cos\Omega t$$

$$\Rightarrow \qquad D_1(t) \;=\; \frac{\frac{U_0}{2}\cos\Omega t - \alpha w(l/2,t)}{g(l/2)}.$$

Then substituting $D_1(t)$ into equation (3.26) yields:

$$\begin{aligned}
\varphi(z,t) &= \alpha w(z,t) + \frac{\frac{U_0}{2}\cos\Omega t - \alpha w(l/2,t)}{g(l/2)}g(z)\\
&= \alpha(w(z,t) - w(l/2,t)f(z)) + \frac{U_0}{2}f(z)\cos\Omega t.
\end{aligned}$$

$\square$

In the previous section, we obtained the partial differential equation (3.8) for $w(z,t)$. This equation can be recast into the form

$$\begin{aligned}
\ddot{w} &= \frac{E^*}{\rho}\left(\frac{1}{y}\cdot(yw')'\right)\\
&=: Lw, \tag{3.27}
\end{aligned}$$

where L is the linear differential operator corresponding with (3.8). Eigenfunctions of L are functions $W(z)$ with the special property that

$$Lw = \lambda w, \tag{3.28}$$

for some number λ. From the theory of partial differential equations it is known that every solution $w(z,t)$ of (3.8) can be expressed as a (possibly infinite) linear combination [Wer02]

$$w(z,t) = \sum_{k=1}^{\infty} W_k(z)p_k(t) \tag{3.29}$$

of the eigenfunctions $W_k(z)$. The problem of solving the partial differential equation (3.8) with the given boundary conditions is then reduced to finding the correct coefficients $p_k(t)$.

When one considers the case of short circuited electrodes of the piezoceramics ($U_0 \equiv 0$), the equation (3.20) can be rewritten as

$$\Phi_k(z) = \alpha(W_k(z) - W_k(l/2)f(z)) \tag{3.30}$$
$$k = 1, 2, \ldots$$

since it is time-independent.

Computation of the eigenfunctions. In the following, we explain how the eigenfunctions $W_k(z)$ are computed as solutions of ordinary differential equations.

The eigenfunctions W_k with eigenfrequencies ω_k^2 are defined by the equation

$$E^* \left(y(z)W_k''(z) + y'(z)W_k'(z) \right) = -\rho y(z)\omega_k^2 W_k(z), \tag{3.31}$$

via the linear differential operator L in equations (3.27) and (3.28).

To solve this equation, we rewrite it as a first order system

$$W_k'(z) = q(z),$$
$$q'(z) = -\frac{y'(z)}{y(z)}W_k'(z) - \frac{\rho\omega_k^2}{E^*}W_k(z). \tag{3.32}$$

Since only shapes with $y(z)$ being even are considered, we can restrict the problem to $z \in [0, l/2]$.

To obtain the boundary conditions of W_k at $z = 0$, we consider the equation (3.30) at $z = -l/2$:

$$\Phi_k(-l/2) = \alpha(W_k(-l/2) - W_k(l/2)f(-l/2))$$

We have already proven that $g(z)$ is odd. Therefore from the definition of $f(z)$ in Proposition (3.2.1), we obtain $f(-l/2) = -1$. The boundary condition (3.6) implies that

$$\Phi(-l/2) = 0,$$

therefore, $W_k(z)$ is an odd function and thus the initial value

$$W_k(0) = 0. \tag{3.33}$$

When $y(z)$ is a constant, we additionally have

$$W_k'(0) = \frac{2\lambda_k}{l}. \tag{3.34}$$

We also use this initial condition in the case of non-constant $y(z)$, with λ_k being determined by solving the characteristic equation

$$(E^0 + \gamma_0 \alpha)\lambda_k \cos \lambda_k - \alpha \gamma_0 \sin \lambda_k = 0 \qquad (3.35)$$

Notice that in equation (3.32) ω_k is still unknown. This is a two-point boundary value problem involving one unknown parameter.

For the short circuit case, the boundary condition (3.5) is

$$E^* W'(l/2) - \frac{\alpha \gamma_0 W(l/2)}{y(l/2)g(l/2)} = 0.$$

We consider this as the third condition when solving the boundary value problem (3.32).

3.2.3   Numerical Solutions

The boundary value problem (3.32) is solved by using the MATLAB function `bvp4c`. Figure 3.2 shows the computation results of $W_1(z)$ and $\Phi_1(z)$ when $y(z) = a|z| + b$. Figure 3.3 shows the computation results of $W_1(z)$ and $\Phi_1(z)$ when $y(z) = az^2 + b$. The red line represents that the shape of a piezo is a rectangle. The green and blue lines represent the shapes with curved sides, where the green line states a convex shape and the blue line states a concave one.

3.3   Derivation of the Equation of Motion for a General Shape

In this section, we first propose an ansatz for $w(z,t)$ and $\varphi(z,t)$. The linear equation of motion of piezoceramics for a general shape is derived via the Calculus of Variations. The nonlinear equation of motion is obtained by introducing two nonlinear terms into the linear equation of motion.

Linear equation of motion.   The virtual work δW in equation (3.1) is not zero in a damping case. For a linear model , we consider the first term of δW in equation (3.2) as a

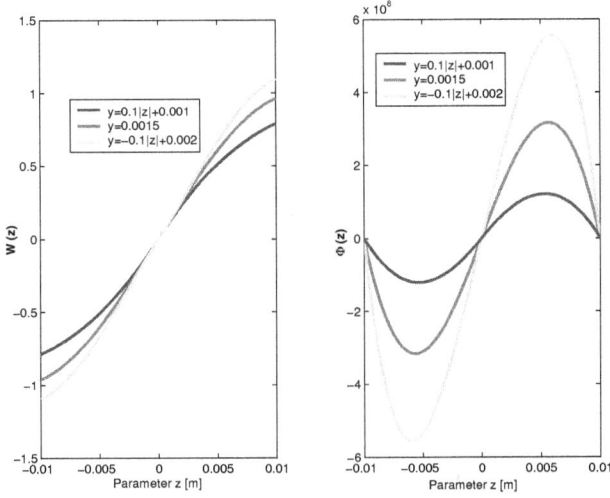

Figure 3.2: Eigenfunctions of the piezo for $y(z) = a|z| + b$

nondimentional damping factor and add it to equation (3.7)

$$\delta \int_{t_0}^{t_1} \int_{-l/2}^{l/2} y(z) \left(\frac{1}{2}\rho\dot{w}^2 - \frac{1}{2}E^0 w'^2 - \gamma_0 w'\varphi' + \frac{1}{2}\nu_0\varphi'^2 \right) dzdt$$

$$- \int_{t_0}^{t_1} \int_{-l/2}^{l/2} y(z) E_d^0 \dot{w}_k' \delta w_k' \, dzdt \;=\; 0. \qquad (3.36)$$

In order to compute the spatial eigenfunctions of a piezo, the following ansatz for $w(z,t)$ and $\varphi(z,t)$ is employed from equations (3.20) and (3.29):

$$w(z,t) \;=\; \sum_{k=1}^{\infty} W_k(z)p_k(t), \qquad (3.37)$$

$$\varphi(z,t) \;=\; \sum_{k=1}^{\infty} \Phi_k(z)p_k(t) + \frac{U_0}{2}f(z)\cos\Omega t. \qquad (3.38)$$

39

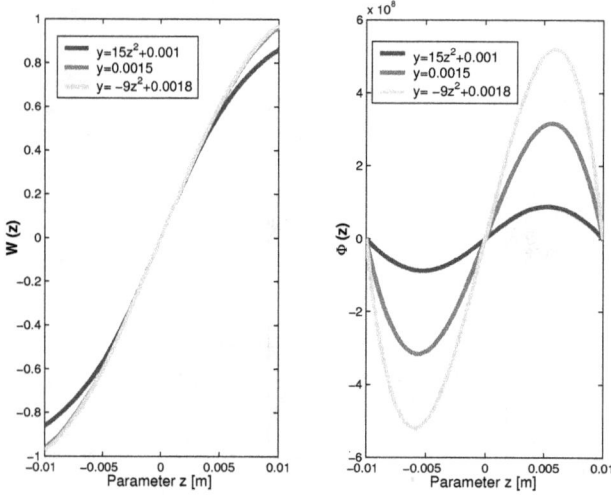

Figure 3.3: Eigenfunctions of the piezo for $y(z) = az^2 + b$

To derive the linear equation of motion, we insert the above ansatz (3.37) and (3.38) into equation (3.36) and obtain

$$\sum_{k=1}^{\infty} \delta \int_{t_0-l/2}^{t_1} \int^{l/2} F(W_k, W_k', p_k, \dot{p}_k)\, dzdt - \sum_{k=1}^{\infty} \int_{t_0-l/2}^{t_1} \int^{l/2} y(z) E_d^0 W_k'^2 \dot{p}_k \delta p_k\, dzdt = 0 \qquad (3.39)$$

with

$$F(W_k, W_k', p_k, \dot{p}_k) = y(z)\Big(\frac{1}{2}\rho W_k^2 \dot{p}_k^{\,2} - \frac{1}{2}E^0 W_k'^2 p_k^2 - \gamma_0 W_k' p_k (\Phi_k' p_k \\ + \frac{U_0}{2}f'(z)\cos\Omega t) + \frac{1}{2}\nu_0(\Phi_k' p_k + \frac{U_0}{2}f'(z)\cos\Omega t)^2\Big)$$

40

and performing the variation with respect to δp_k

$$F_{p_k} - \frac{\partial}{\partial t} F_{\dot{p}_k} - \delta W_{p_k} = 0$$

$$\Rightarrow \int_{-l/2}^{l/2} \Big(-E^0 W_k'^2 y p_k + \nu_0 \Phi_k'^2 y p_k - 2\gamma_0 W_k' \Phi_k' y p_k$$

$$- \frac{U_0}{2}\gamma_0 W_k' f'(z) y \cos\Omega t + \frac{U_0}{2}\nu_0 \Phi_k' f'(z) y \cos\Omega t$$

$$- \frac{\partial}{\partial t}\rho W_k^2 y \dot{p}_k \Big) dz - \int_{-l/2}^{l/2} \big(y E_d^0 W_k'^2 \dot{p}_k \big) dz = 0$$

$$\Rightarrow \int_{-l/2}^{l/2} \Big(y\rho W_k^2 \ddot{p}_k + (E^0 W_k'^2 + 2\gamma_0 W_k' \Phi_k' - \nu_0 \Phi_k'^2) y p_k$$

$$+ y E_d^0 W_k'^2 \dot{p}_k + \frac{U_0}{2} y f'(z)\cos\Omega t(\gamma_0 W_k' - \nu_0 \Phi_k') \Big) dz = 0. \tag{3.40}$$

We already defined that

$$f(z) = \frac{g(z)}{g(\frac{1}{2})}$$

and

$$g(z) = \int \frac{1}{y(z)} dz.$$

Therefore

$$f'(z) = \frac{1}{y(z)g(\frac{1}{2})}.$$

Then equation (3.40) can be simplified to

$$m_k \ddot{p}_k + d_k \dot{p}_k + c_k p_k = f_k \cos\Omega t, \tag{3.41}$$

$$k = 1, 2, \ldots$$

where

$$m_k = \rho \int_{-l/2}^{l/2} y W_k^2 dz, \quad d_k = E_d^0 \int_{-l/2}^{l/2} y W_k'^2 dz,$$

$$c_k = E^0 \int_{-l/2}^{l/2} y W_k'^2 dz + 2\gamma_0 \int_{-l/2}^{l/2} y W_k' \Phi_k' dz - \nu_0 \int_{-l/2}^{l/2} y \Phi_k'^2 dz,$$

$$f_k = -\gamma_0 \frac{U_0}{2g(l/2)} \int_{-l/2}^{l/2} W_k' dz + \nu_0 \frac{U_0}{2g(l/2)} \int_{-l/2}^{l/2} \Phi_k' dz,$$

41

The coefficients ρ, E^0, γ_0, E_d^0 and ν_0 in m_k, c_k, d_k and f_k are constants. d_k is a nondimensional damping factor.

Nonlinear equation of motion.   There are many possibilities to describe a nonlinear phenomenon with different models. In [Neu02], a nonlinear model with many nonlinear terms is introduced. It also indicated and tested that the nonlinear phenomenon observed in experiments can be sufficiently described with only two terms. With three or more nonlinear terms, the results are no more accurate but the computation will be much more complicated.

In Neumann's work [Neu02], several combination pairs of the nonlinear terms are analyzed. Here we introduce a nonlinear equation of motion with two nonlinear terms. One nonlinear term is from equation (3.4) as

$$\delta \int_{t_0}^{t_1} \int_{-l/2}^{l/2} -\frac{1}{4} y E^2 w'^4 \, dz dt,$$

and the other is a nondimensional damping term from δW as

$$\int_{t_0}^{t_1} \int_{-l/2}^{l/2} -y E_d^2 \dot{w}_k'^3 \delta w_k' \, dz dt.$$

Adding these two nonlinear terms into the linear equation of motion (3.39) we obtain

$$\sum_{k=1}^{\infty} \delta \int_{t_0}^{t_1} \int_{-l/2}^{l/2} F(W_k, W_k', p_k, \dot{p}_k) \, dz dt - \sum_{k=1}^{\infty} \int_{t_0}^{t_1} \int_{-l/2}^{l/2} y E_d^0 W_k'^2 \dot{p}_k \delta p_k \, dz dt$$

$$- \sum_{k=1}^{\infty} \delta \int_{t_0}^{t_1} \int_{-l/2}^{l/2} \frac{1}{4} y E^2 W_k'^4 p_k^4 \, dz dt - \sum_{k=1}^{\infty} \int_{t_0}^{t_1} \int_{-l/2}^{l/2} 3y E_d^2 W_k'^4 p_k^2 \dot{p}_k \delta p_k \, dz dt \quad = 0 \quad (3.42)$$

Performing the variation with respect to δp_k on the two newly added nonlinear parts we obtain

$$\frac{\partial}{\partial p_k} \int_{t_0}^{t_1} \int_{-l/2}^{l/2} \frac{1}{4} y E^2 W_k'^4 p_k^4 \, dz dt + \int_{-l/2}^{l/2} 3 E_d^2 W_k'^4 p_k^2 \dot{p}_k \, dz$$

$$= \int_{-l/2}^{l/2} y E^2 W_k'^4 p_k^3 \, dz + \int_{-l/2}^{l/2} 3 E_d^2 W_k'^4 p_k^2 \dot{p}_k \, dz$$

$$(3.43)$$

Then, we obtain the following nonlinear equation of motion

$$m_k \ddot{p}_k + d_k \dot{p}_k + c_k p_k + \varepsilon_k p_k^3 + \varepsilon_{d_k} p_k^2 \dot{p}_k = f_k \cos \Omega t, \qquad (3.44)$$

$$k = 1, 2, \ldots$$

with

$$\varepsilon_k = E^2 \int_{-l/2}^{l/2} y W_k'^4 dz,$$

$$\varepsilon_{d_k} = 3E_d^2 \int_{-l/2}^{l/2} y W_k'^4 dz,$$

where the constant E^2 is obtained experimentally. m_k, c_k and f_k are the same as in the linear equation of motion. The influence of the two nonlinear terms ε_k and ε_{d_k} on the nonlinearities will be analyzed in Section 4.3.

Simplification of the ansatz and extension to a nonlinear model for a general shape.
In this part, the given ansatz is first simplified for a rectangular piezo with a linear model by using a Galerkin method. Then, we assume that the simplification of the ansatz is also available for a curved side shape as well as for a nonlinear model.
For a rectangular piezo, the special case of $y(z)$ being a constant is considered. Therefore in equation (3.38)

$$f(z) = \frac{2z}{l}.$$

and the equation (3.8) can be simplified to:

$$E^* w''(z, t) = \rho \ddot{w}(z, t), \qquad (3.45)$$

which can be solved analytically. The eigenfunctions of the linear problem with short circuited electrodes ($U_0 \equiv 0$) are calculated here in order to obtain suitable shape functions.

The eigenfunctions are obtained by solving equation (3.45) with the boundary conditions analytically [vWH03]:

$$W_k(z) \;=\; \sin(\lambda_k \frac{2z}{l}), \tag{3.46}$$

$$\Phi_k(z) \;=\; \alpha(W_k(z) - \frac{2z}{l}\sin(\lambda_k)), \tag{3.47}$$

with

$$\lambda_k^2 = \frac{l^2 \rho \omega_k^2}{4 E^*},$$

where ω_k is the kth circular eigenfrequency, and λ_k can be determined by solving equation (3.35).

We already derived the linear equation of motion (3.41) for a general shape above. For a rectangular shape, the coefficients m_k, d_k, c_k and f_k can be simplified as

$$m_k \;=\; \rho \int_{-l/2}^{l/2} W_k^2 dz, \quad d_k = E_d^0 \int_{-l/2}^{l/2} W_k'^2 dz,$$

$$c_k \;=\; E^0 \int_{-l/2}^{l/2} W_k'^2 dz + 2\gamma_0 \int_{-l/2}^{l/2} W_k' \Phi_k' dz - \nu_0 \int_{-l/2}^{l/2} \Phi_k'^2 dz,$$

$$f_k \;=\; -\gamma_0 \frac{U_0}{l} \int_{-l/2}^{l/2} W_k' dz,$$

A Galerkin approach The numerical results of p_k under certain excitation frequencies can be obtained by solving equation (3.41) numerically ($U_0 \neq 0$) with MATLAB function ode45 for $k = 1, 2, \cdots$. Here, we first compute the eigenfunctions W_k via equation (3.46) from the first to fourth modes (e.g. $k = 1, 2, 3, 4$ respectively). We then calculate the corresponding m_k, c_k, d_k and f_k above and solve the equation (3.41) when the excitation frequency Ω is at the first resonance numerically. Finally we obtain the numerical results of p_k and thus can compute the $w(z)$ in equation (3.37) for different k. Figure 3.4 shows the computation of equation (3.37) for $k = 1, 4$ respectively. Via equation (3.46) we could obtain the analytical solutions of W_k for different z, and the p_k are

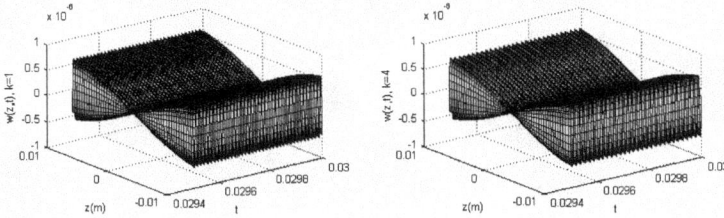

Figure 3.4: Plotting of $w(z,t) = \sum_{i=1}^{k} W_i(z)p_i(t), k = 1, 4$ respectively

obtained by solving equation (3.41) numerically when $U_0 = 20V$. It's difficult to clearly see the difference between the two charts in Figure 3.4.

Then we plot not the sum but the $W_k(z)p_k(t)$ for $k = 1, 2, 3, 4$ respectively in figure 3.5. When $k = 1$, the power of the vibration amplitude for the first eigenmode is 10^{-6}, and when $k = 2, 3, 4$, the powers are 10^{-10} and 10^{-11} which are much smaller than that of the first eigenmode. From the analysis of the first four eigenmodes, the results reveal that the second to fourth eigenmodes show very small contributions to the vibration compared to the first when the excitation frequency is close to the first resonance. Therefore we could obtain sufficiently exact results by considering only the first eigenmode. Consequently in the following chapters the subscript k will be omitted.

Bearing this in mind, the ansatz is simplified to

$$w(z,t) = W(z)p(t), \qquad (3.48)$$

$$\varphi(z,t) = \Phi(z)p(t) + \frac{U_0}{2}f(z)\cos\Omega t, \qquad (3.49)$$

for the system excited close to the first resonance frequency and the subscript 1 is omitted.

Assumption for a nonlinear model of a general shape. For a linear model, we already proved above that the ansatz (3.37) and (3.38) can be simplified as (3.48) and (3.49) for a rectangular shape. Howeever, we could not prove that the simplification is also available for a nonlinear model. Since engineers have already used the simplified ansatz

45

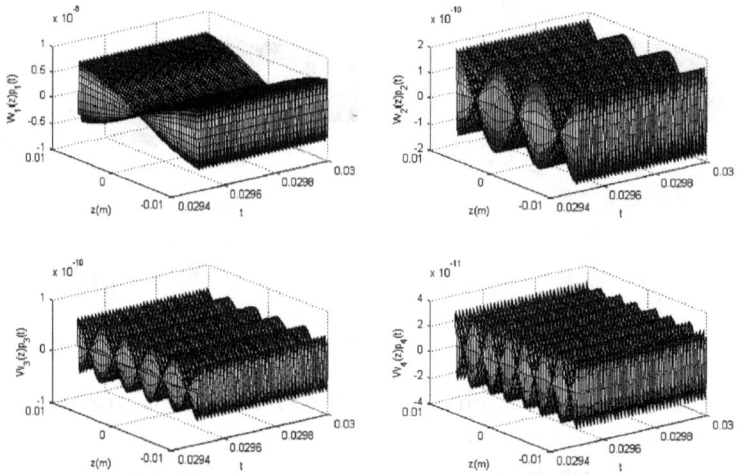

Figure 3.5: Plotting of $W_k(z)p_k(t), k = 1, 2, 3, 4$ respectively

for a nonlinear model [vWH03], we assume that the simplification can also be used for a nonlinear model as well as a general shape (e.g a curved side shape).

Chapter 4

Solutions of Equations of Motion and Nonlinear Dynamical Model Analysis

In Sections 4.1 and 4.2, the linear and nonlinear equations of motion are solved numerically via a continuation software package AUTO2000, and the results show that some curved side piezoceramics have a perform better than those with a rectangular shape. In Section 4.3, the effects of the two nonlinear terms are analyzed, and both nonlinear terms have profound effects on the bifurcation.

4.1 Numerical Solutions for a Linear Model

In Section 3.3 the ansatz is simplified as (3.48) and (3.49). Therefore, the linear equation of motion (3.41) for a general shape can also be simplified as

$$m\ddot{p} + d\dot{p} + cp = f \cos \Omega t, \tag{4.1}$$

where

$$
m = \rho \int_{-l/2}^{l/2} y W^2 dz, \quad d = E_d^0 \int_{-l/2}^{l/2} y W'^2 dz
$$

$$
c = E^0 \int_{-l/2}^{l/2} y W'^2 dz + 2\gamma_0 \int_{-l/2}^{l/2} y W' \Phi' dz - \nu_0 \int_{-l/2}^{l/2} y \Phi'^2 dz,
$$

$$
f = \gamma_0 \frac{U_0}{2g(l/2)} \int_{-l/2}^{l/2} W' dz - \nu_0 \frac{U_0}{2g(l/2)} \int_{-l/2}^{l/2} \Phi' dz.
$$

The physical values are:

$$
\rho = 7850 \ [\frac{kg}{m^3}]; \qquad E^0 = 7.0912 \cdot 10^{10} \ [\frac{N}{m^2}];
$$

$$
\gamma_0 = 16.1608 \ [\frac{N}{Vm}]; \qquad \nu_0 = 6.3665 \cdot 10^{-9} \ [\frac{Nm^2}{V^3}];
$$

$$
E_d^0 = 120 \ [\frac{Ns}{m^2}];
$$

We rewrite equation (4.1) as:

$$
\dot{p} = q,
$$

$$
\dot{q} = -\frac{1}{m}(dq + cp - f\cos(\Omega t)).
$$

This is a first order system for which periodic solutions can be computed by using the continuation package AUTO2000 [Dea01].

Figure 4.1 shows the periodic solutions of the equation (4.1) when $y(z)$ is a constant $0.0015 \ [m]$. The corresponding coefficients in equation (4.1) are computed and their values are:

$$
m = 0.0939,
$$

$$
d = 36.2940,
$$

$$
c = 2.246 \cdot 10^{10},
$$

$$
f = -46.6503.
$$

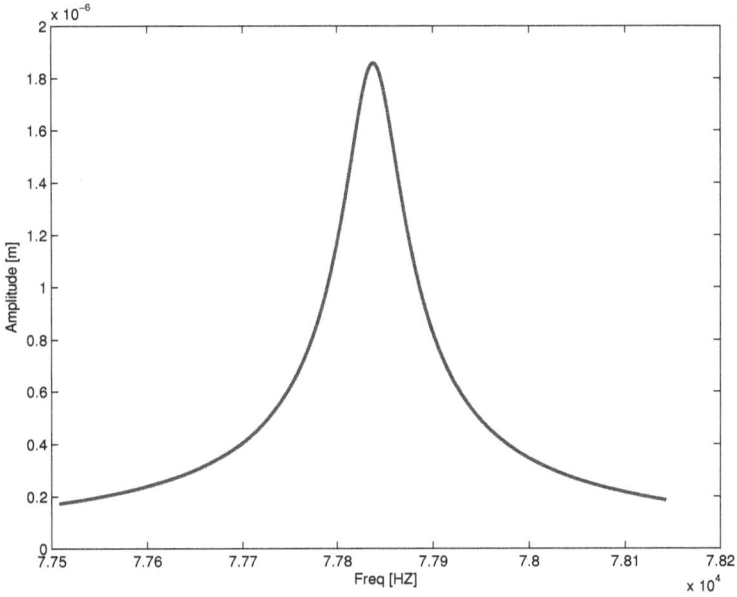

Figure 4.1: The periodic solutions of the linear equation of motion for the rectangular piezo

In the figures in this chapter, the horizontal axis represents the excitation frequency Ω, and the vertical axis represents vibration amplitude calculated using the scaled L_2-norm $(|v| = \{\frac{1}{T}\int_0^T v(t)^2 dt\}^{\frac{1}{2}})$.

For a curved side piezo (e. g. $y(z) = 15z^2 + 0.001$), we recomputed the above coefficients as:

$$
\begin{aligned}
m &= 0.1048, \\
d &= 26.2195, \\
c &= 1.5886 \cdot 10^{10}, \\
f &= -38.8388,
\end{aligned}
$$

49

and replotted the results for rectangular and curved piezo together into Figure 4.2.

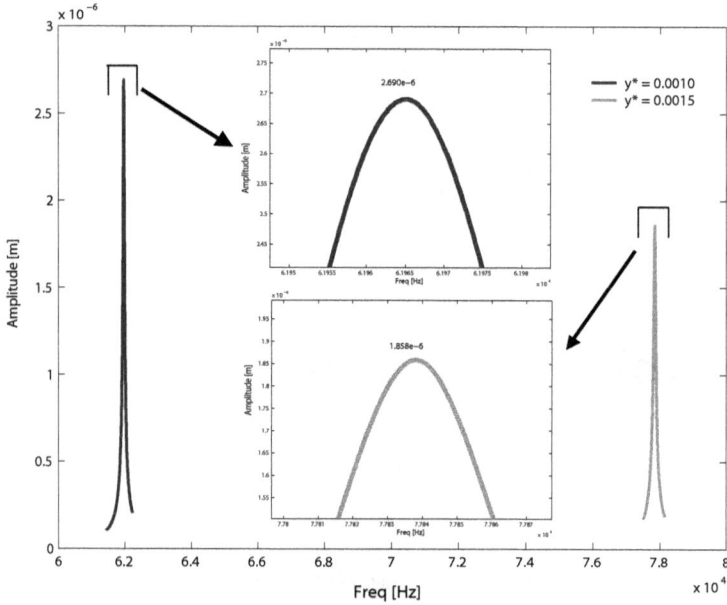

Figure 4.2: Linear model: oscillation amplitude of the piezo dependent on the excitation frequency for different geometries: rectangular (green) and curved shape (blue).

Figure 4.2 shows the vibration amplitudes for two different geometries dependent on the excitation frequency. The green line represents the behavior of a rectangular piezo (i.e. $y(z) = y^*$ is constant), the maximum amplitude at the resonance frequency is $1.858 \cdot 10^{-6}$ m. The blue line represents the piezo with the curved side $y(z) = 15z^2 + 0.001$. The corresponding maximum amplitude at resonance is $2.690 \cdot 10^{-6}$ m, an improvement of about 45%.

4.2 Numerical Solutions for a Nonlinear Model

For the nonlinear equation of motion, we do the same simplification to equation (3.44)

$$m\ddot{p} + d\dot{p} + cp + \varepsilon p^3 + \varepsilon_d p^2 \dot{p} = f \cos \Omega t, \tag{4.2}$$

where

$$\varepsilon = E^2 \int_{-l/2}^{l/2} y W'^4 dz,$$

$$\varepsilon_d = 3E_d^2 \int_{-l/2}^{l/2} y W'^4 dz,$$

and constants E^2 and E_d^2 are also obtained experimentally and their values are:

$$E^2 = -4.1820 \cdot 10^{16} \quad [\frac{N}{m^2}]; \quad E_d^2 = 3.0050 \cdot 10^9 \quad [\frac{Ns}{m^2}]$$

For the rectangular piezo y is a constant, the two nonlinear terms are computed and their values are

$$\varepsilon = -1.6041 \cdot 10^{20},$$

$$\varepsilon_d = 3.458 \cdot 10^{13}.$$

m, c, d and f are the same as in the linear case.

The equation (4.2) can be rewritten to:

$$\dot{p} = q,$$

$$\dot{q} = -\frac{1}{m}(dq + cp - f\cos(\Omega t) + \varepsilon p^3 + \varepsilon_d p^2 q).$$

The computation for periodic solutions is also done using AUTO2000. The difference between the solutions for linear and nonlinear models is that there are unstable solutions and two limit points in the nonlinear case because of the nonlinear effects of the two nonlinear terms ε and ε_d.

Figure 4.3 shows that in the case of a rectangular piezo ($y(z) = y^* = 0.0015$ m), for certain excitation frequencies there exist *three* periodic solutions, two of which (indicated by the solid lines) are stable and one of which is unstable (dashed line) and therefore cannot be observed in experiments. This means that the periodic solutions undergo two bifurcations at certain frequencies – in fact, two limit points exist. In particular, this result explains the "jump phenomenon" observed in the behavior of the nonlinear model in [vWH03].

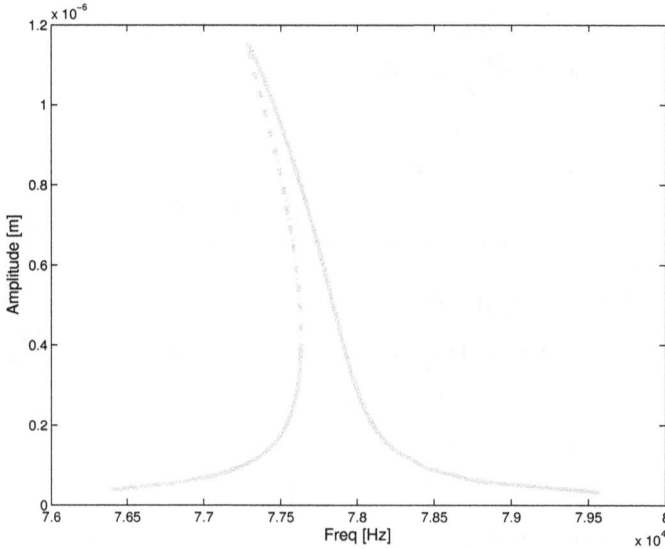

Figure 4.3: Nonlinear model: path of periodic solutions within a certain range of excitation frequencies (piezo with rectangular shape).

In order to compare the dynamical behavior for different shapes of the piezo, we focus on the right limit points (at Ω_1 and Ω_2 respectively in Figure 4.4) instead of the resonance frequencies, since the piezo may show unstable behavior in the latter case. In Figure 4.4, the green line represents the piezo with rectangular shape, and the blue line represents the curved shaped piezo. For the curved side $y(z) = 15z^2 + 0.001$, the corresponding

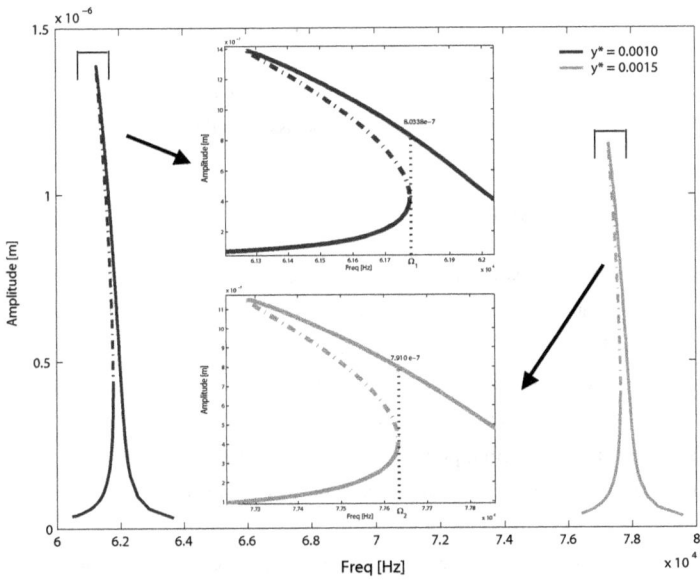

Figure 4.4: Nonlinear model: paths of periodic solutions within a certain range of excitation frequencies for different geometries of the piezo: rectangular (green) and curved shapes (blue).

nonlinear terms are computed

$$\varepsilon = -1.2228 \cdot 10^{20},$$
$$\varepsilon_d = 2.6359 \cdot 10^{13}$$

Comparing the amplitudes at the chosen points for different geometries, we observe that for the rectangular piezo, the amplitude at $\Omega_2 = 77632$ Hz is $7.91 \cdot 10^{-7}$ m, while the amplitude at $\Omega_1 = 61776$ Hz for the curved piezo is $8.34 \cdot 10^{-7}$ m, which is an improvement of more than 5%. This result again indicates that one might be able to improve the performance of a certain actuator by changing the shape of the associated piezo.

4.3   Nonlinear Effects Analysis

In the previous section, we fixed the parameters ε and ε_d of the nonlinear model (3.44). We now keep $y(z) \equiv y^*$ constant and study the effect of varying these parameters on the computed branches of periodic solutions.

We first vary ε, keeping ε_d fixed. For ε beyond a certain threshold value (corresponding to the yellow line in Figure 4.5), no limit point exists and the system is stable. As ε decreases (that is, as $|\varepsilon|$ increases), two limit points appear that separate frequencies with only one stable solution from a range of frequencies for which there are two stable and one unstable periodic solution.

In Figure 4.6, ε is fixed and ε_d is varied. For large ε_d there are no limit points (the yellow and the cyan line in Figure 4.6). As ε_d decreases, the maximum amplitude increases and the two limit points appear.

From the above results, we see that both ε and ε_d have profound effects on the bifurcation diagram of the system. In addition the influence of the value of ε_d on amplitude is also observed, as it comes from the virtual work δW (see equation 3.2) and thus plays also a damping role on the system.

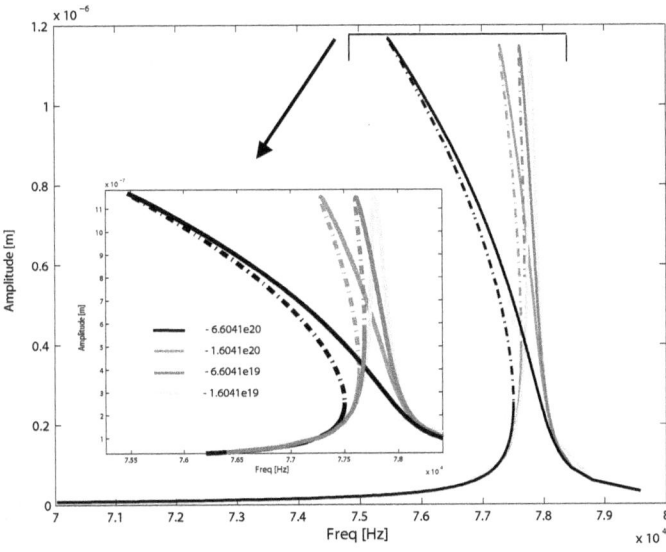

Figure 4.5: Nonlinear model: paths of periodic solutions within a certain range of excitation frequencies for different values of the parameter ε.

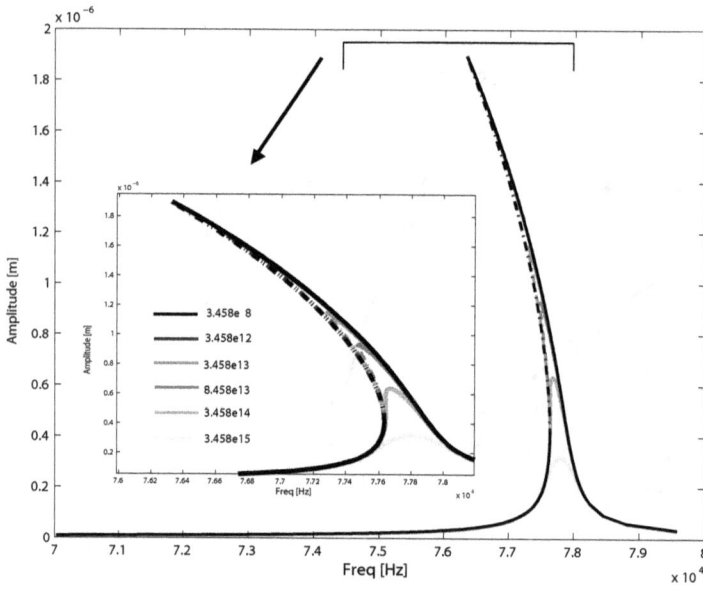

Figure 4.6: Nonlinear model: paths of periodic solutions within a certain range of excitation frequencies for different values of the parameter ε_d.

Chapter 5

Multi-Objective Optimization Problems (MOOPs)

In this chapter, the basic concepts used in multi-objective optimization problems (MOOPs) are formally defined, and the principles of multi-objective optimization are outlined in Section 5.1. In Section 5.2 the classification and an overview of the main optimization techniques for MOOPs currently available are presented, and a few corresponding software packages are listed. Particularly, a Set Oriented Multilevel Subdivision Technique is studied in Section 5.3, as it plays an important role in this dissertation. In Section 5.4, a multi-objective shape optimization problem for piezoceramics is introduced, and a GAIO (Global Analysis of Invariant Objects) model for solving the optimization problem is proposed.

5.1   A Brief Introduction to MOOPs

Almost every real-world problem naturally involves simultaneous optimization of several incommensurable objectives. For example, when we design a chemical process, we will normally want to maximize its economic performance (including profit, fixed cost, operation and maintenance cost, etc), but at the same time, we incorporate several objectives involving environmental sustainability, safety, operability, and controllability. The objectives normally conflict with each other. These problems are called multi-objective or vector optimization problems. A considerable number of researchers have produced a number of theoretical and practical contributions to deal with MOOPs over the past five decades. Current applications of multi-objective optimization are distributed widely in engineering, industrial and scientific fields. Good reviews of the techniques for multi-objective optimization can be found in the books ([Mie99], [Deb01], [Ehr02]). These publications provide a very broad bibliography in this area.

5.1.1 Formulation of the Multi-objective Optimization Problem

To formulate a multi-objective optimization problem, we first define

$$x = [x_1, x_2, \ldots, x_n]^T$$

as decision variables which subject to m inequality constraints:

$$g_i(x) \leq 0 \quad i = 1, 2, \ldots, m \tag{5.1}$$

and p equality constraints:

$$h_i(x) = 0 \quad i = 1, 2, \ldots, p. \tag{5.2}$$

We also define $\mathcal{X} \in \mathbf{R}^n$ as a set of x satisfying the constraints, then put all objective functions f_i ($i = 1, \ldots, k$) into one function

$$F : \mathbf{R}^n \to \mathbf{R}^k.$$

where

$$F(x) = [f_1(x), \ f_2(x), \ldots, f_k(x)]^T, \tag{5.3}$$

k is the number of objective functions
From the set $\mathcal{X}$ we wish to determine the particular of values $x_1^*, x_2^*, \ldots, x_n^*$ which yield the optimal values for all of the objective functions.

5.1.2 Pareto Optimality and Pareto Set

While in a single-objective optimization problem a well-defined optimal solution can normally be found, seldomly there is a single optimum that simultaneously minimizes all the objective functions for MOOPs. Instead of a single optimum we can find a set of trade-off solutions. The concept of Pareto optimality is used to deal with this case[CC05].

Definition 5.1. *We say that a vector of decision variables $x^* \in X$ is Pareto optimal if there does not exist another $x \in X$ such that $f_i(x) \leq f_i(x^*)$ for all $i = 1, \ldots, k$ and $f_j(x) < f_j(x^*)$ for at least one j.*

In words, this definition says that x^* is Pareto optimal if no feasible vector of decision variables $x \in X$ exists which would decrease some criterion without causing a simultaneous increase in at least one other criterion. Unfortunately, this concept almost always results in not a single solution, but rather a set of solutions called the *Pareto set*. The vectors x^* corresponding to the solutions included in the Pareto set are called *nondominated*. The image of the Pareto set under the objective functions is called *Pareto front*.

5.2 Methods to Solve MOOPs

Finding the entire Pareto set is the most important step of solving a multi-objective optimization problem. In this section, different methods for solving MOOPs are reviewed, and the corresponding software packages and literature are listed.

5.2.1 An Overview

A variety of powerful techniques for solving MOOPs has resulted from operations research, engineering, computer science and other related disciplines for years.
A considerable overview of deterministic techniques is given by Miettinen [Mie99]. Earlier years, MOOPs were often solved by traditional optimization techniques. These techniques usually reduce the MOOP to a single objective optimization problem, either by combining multiple objective into a single scalar function or by keeping one of the objectives and restricting the rest of the objectives with user-specified values. While these methods, such as weighted sum method, ε-constraint method, goal programming method, interactive methods, the min-max approach and so on, are usually easy to implement and to use, they have several disadvantages:

1. All techniques may miss some optimal solutions.

2. They may depend on the shape of the search space, e.g. whether it is convex or not.

3. They are time-consuming methods because it is necessary to do a series of separate optimization runs to obtain the Pareto set.

4. All techniques require some problem knowledge, such as suitable weights, ε or target values.

Because many MOOPs are high dimensional, discontinuous, multi-modal and/or Non-deterministic Polynomial (NP)-Complete, stochastic methods often yield better performance.

Multi-objective Evolutionary Algorithms (MOEA) represent a stochastic technique inspired by the principles of natural selection and natural genetics. Such techniques have been demonstrated to be very powerful and suitable for solving MOOPs because of their ability to find the Pareto set. Most researchers on MOEA have concentrated their efforts in developing new and efficient search algorithms for finding widely distributed Pareto set. Reviews of different evolutionary approaches to multi-objective optimization have been given by researchers ([FF95] , [Hor97] , [vVL00], [CC05], [Zit99], [Deb01] and [Tea01]). The classification of approaches that follows is partially based on the discussion presented in these references.

5.2.2   Software

Table 5.1 lists some software packages for solving MOOPs.

5.3   A Set Oriented Multilevel Subdivision Technique

A set oriented numerical method for the numerical solution of multi-objective optimization problems is introduced in this section. This method is global in nature and permits an approximation of the entire set of global Pareto points.

5.3.1   A Set Oriented Multilevel Subdivision Technique

Dellnitz and Hohmann [DH97] proposed a subdivision algorithm for the computation of unstable manifolds and global attractors. Suppose that the dynamical system is defined

Table 5.1: Some software packages for MOOPs

Software	Developers	Techniques Used	Reference
AMOPSO	G. T. Pulido	Particle Swarm Optimizer	[PC04]
ε-MOEA	K. Deb	Evolutionary Algorithm	[Deb]
GAIO	M. Dellnitz & O. Junge	Multilevel Subdivision Techniques	[DFJ01]
MOEA toolbox	K. C. Tan	Evolutionary Algorithm	[Tea01]
MOMHLib++	A. Jaszkiewicz	Evolutionary Algorithm Simulated Annealing	[Jas]
NIMBUS	K. Miettinen	Interactive NIMBUS Method	[MM00]
NSGA-II	K. Deb	Evolutionary Algorithm	[Dea02]
PAES	J. D. Knowles	Evolutionary Algorithm	[KC00]
PISA	Eckart Zitzler	Evolutionary Algorithm	[BLTZ03]

on R^n. First, specify and subdivide a box in R^n and throw away boxes which do not contain part of the relative global attractor. Then, derive the boxes again and proceed in the same manner. For a detailed explanation on the subdivision algorithm refer to [DH97] and [DSS02].

The idea of the new set oriented numerical method is to write down an iteration scheme which - interpreted as a discrete dynamical system - possesses the Pareto set as an attractor. Then set oriented numerical methods for dynamical systems can be used for its approximation.

More concretely three different set oriented multilevel approaches for the approximation of the Pareto set are proposed in [DSH05]. First a *subdivision algorithm* for the approximation of Pareto sets which creates tight box coverings of these objects is presented. The second algorithm is a *recovering algorithm* which can be viewed as a postprocessing procedure for the subdivision scheme yields the second approach. In the third approach the creation of the box covering is combined with appropriate branch and bound strategies by a *sampling algorithm*. More details can be found in [Sch04].

Subdivision Algorithm.   The subdivision algorithm is directly based on the theoretical considerations of the work [DSH05], particularly on Corollaries 1 and 2.

For a finite collection of discrete dynamical systems, the initial value problem

$$\dot{x}(t) = -q(x(t)), \quad x(0) = 0 \tag{5.4}$$

is discretized and the following iteration scheme

$$x_{j+1} = x_j + h_j p_j, \quad j = 0, 1, 2, \ldots, \tag{5.5}$$

is considered.

Corollary 1. *Suppose that the set $\mathcal{S}$ of substationary points is bounded and let $\mathcal{D}$ be a compact neighborhood of $\mathcal{S}$. Then an application of the subdivision algorithm to $\mathcal{D}$ with respect to iteration scheme (5.5) creates a covering of the entire set $\mathcal{S}$, that is,*

$$\mathcal{S} \subset Q_k \quad for \quad k = 0, 1, 2, \ldots,$$

in the course of the subdivision process.

Corollary 2. *Suppose that the set $\mathcal{S}$ of substationary points is bounded and connected. Let $\mathcal{D}$ be a compact neighborhood of $\mathcal{S}$. Then an application of the subdivision algorithm to $\mathcal{D}$ with respect to the iteration scheme (5.5) leads to a sequence of covering which converges to the entire set $\mathcal{S}$, that is,*

$$h(\mathcal{S}, Q_k) \to 0 \quad for \quad k = 0, 1, 2, \ldots,$$

where h denotes the usual Hausdorff distance.

The descent direction used in the computation of unconstrained MOOPs is $p_j = q(x_j)$. A particular Armijo step size strategy is chosen in the following way: starting with the given points x_j, F is evaluated along the descent direction p_j in uniform step lengths h_0 as long as the value of all objectives decreases. Once one objective function starts to increase, a "better" iterate x_{j+1} with intermediate step length is calculated via backtracking.

The *subdivision algorithm* has the advantage of being very robust with respect to errors by the use of the descent direction. However, all the gradients of the objectives have to be available and the algorithm is unable to distinguish between a local and a global Pareto point. Furthermore, the efficiency of the algorithm will get worse when the MOOP has optima relative to the boundary of the domain.

Recovering Algorithm. It may be the case that in the course of the subdivision procedure boxes get lost although they contain substationary points. This will, for instance, be the case when there are not enough test points taken into account for the evaluation of $\mathbf{F}(B)$ for a box $B \in \mathcal{B}_k$. The recovering algorithm uses a kind of "healing" process which allows us to recover those substationary points which have previously been lost.

The aim of the algorithm is to extend the given box collection step-by-step along the covered parts of the set $\mathcal{S}$ of the substationary points no more boxes are added.

The *recovering algorithm* is able to extend the computed box covering of the set of substationary points but it is only local in nature.

Sampling Algorithm. Observe that there are a few potential drawbacks which may occur when using the two algorithms described above:

1. the gradients of the objectives are needed,

2. the set $\mathcal{S}$ is generally a strict superset of the Pareto set, and

3. the algorithms are capable of finding local Pareto points on the boundary of the domain Q - e.g. via penalization strategies. However, it has turned out in practice that (MOOPs) typically contain many local Pareto points on ∂Q which are not globally optimal (see e.g. in [DSH05]).

The sampling algorithm avoids all these problems because it takes only the function values of the objective functions into account. On the other hand this algorithm is not as robust to errors as the first two because it is only global relative to the underlying box collection. The *sampling algorithm* is able to detect global Pareto points even on the boundary of the domain due to the fact that it works in the image space of the MOOP. Naturally, uncertainty always remains due to the sampling approach, in particular when the boxes are big and/or the dimensions of the MOOP are large. Nevertheless, results have shown that this algorithm works quite well, in particular when the gradients of the objectives are not available and the dimension of the MOOP is moderate [Sch04].

5.3.2 Software

GAIO [1] is developed by Dellnitz and Junge [DFJ01]. It is a software package for the global numerical analysis of dynamical systems and optimization problems based on set oriented techniques. It may be used to compute invariant sets, invariant manifolds, invariant measures and almost invariant sets in dynamical systems and to compute the globally optimal solutions of both scalar and multi-objective problems.

5.4 A GAIO Model for the Multi-Objective Optimization Problem

In this section, we will introduce a multi-objective shape optimization problem for the design of piezoelectric actuators. In Section 5.4.1, multi-objective optimization problems in piezoelectric actuator design are introduced. Particularly, a multi-objective shape optimization problem is presented in Section 5.4.2; the objectives, design variables and constraints are given in detail. A GAIO model for the multi-objective optimization problem is proposed in Section 5.4.3.

5.4.1 MOOPs in Piezoelectric Actuator Design

Piezoelectric actuators have been widely used in different fields. The performance of piezoelectric actuators can usually be remarkably improved if mathematical optimization methods are applied in their development. According to the different applications, different design goals exist. For example, the fast response time and low power consumption are considered in one stroke driving; and low cost and compact size are considered in resonant driving.

Mathematical analysis of an optimization problem often leads to "unusual" solutions that are hardly suitable for manufacturing. This is acceptable in the framework of the chosen approach:

---

[1] http://www-math.upb.de/ agdellnitz/Software/gaio.html

We are looking for a mathematically correct solution, and we accept its features. From a practical point of view, the emergence of "strange" solutions reveals certain hidden features of optimality. These solutions should not be rejected as mathematical extravagance, but rather should be understood and interpreted in depth; often, they point to better solutions that may be approximated with available resources.

C. Cherkaev

The above quotation can also be applied to interpret the results we obtained in Chapter 4. Those results show that one can get better performance (e.g. amplitude) with unusual shapes. However from an engineer's point of view, unusual shapes often cause manufactural problems. This conflict leads to the multi-objective optimization problem in this work.

5.4.2   The Multi-objective Shape Optimization Problem

This work is concerned with a multi-objective shape optimization problem of piezoelectric materials. A two-dimensional body (see the given example shape in Figure 3.1) is to be designed for maximum amplitude (better performance) and minimum curvature (simple manufacturing) subjected to three constraints.

Parametrization of shapes.   To realize a numerical shape optimization problem one has to first find a suitable parametrization of shapes using a finite number of parameters. Concretely we consider two cases to represent the boundary $y(z)$.

- **Design variable.**

    1. $y(z) = az^2 + b$ (one design variable);

       To simplify the problem, we use one design variable to control the shape of piezoceramics. Here we define an admissible domain Q. The shape of the piezo is characterized solely by $y^* \in Q$. $y^* = y(0)$ is the design variable (see Figure 3.1). Together with the mass constraint, it determines the shape of the piezo in a unique way.

65

2. B-spline (two design variables).

From a mathematical point of view, a curve generated by using the vertices of a control polygon is dependent on some interpolation or approximation scheme to establish the relationship between the curve and the control polygon. The scheme is provided by the choice of the basis function.

Curve representation. Curves are mathematically represented either explicitly, implicitly or parametrically. Explicit representations of the form $y = f(x)$ (e.g. Case 1) are useful in many applications but axis-dependent, cannot adequately represent multiple-valued functions and cannot be used where a constraint involves an infinite derivative. Implicit representations of the form $f(x, y) = 0$ for curves are capable of representing multiple-valued functions but are still axis-dependent.

Parametric curve representations of form

$$x = f(t), \qquad y = g(t),$$

where t is the parameter, are extremely flexible. They are axis independent, easily represent multiple-valued functions and infinite derivatives, and have additional degrees of freedom compared to either explicit or implicit formulations. The derivatives of y and x with respect to t are given by

$$y' = \frac{dy}{dt}, \quad x' = \frac{dx}{dt}.$$

B-spline curves we will use in this work are parametrically represented by [Rog01]:

$$p(t) = (x(t), y(t))^T.$$

B-spline definition. A B-spline is defined by a knot vector $K_m = [k_0, k_1, \ldots, k_m]$, where K_m is a nondecreasing sequence, and given control points $B_0, B_1, \ldots, B_n \in \mathbb{R}^2$.

The degree d and the order r are defined as:

$$d = m - n - 1$$
$$r = m - n.$$

The "knots" $k_{d+1}, \ldots, k_{m-d-1}$ are called internal knots.

The dash-dot line defined by the control points will be called a *control polygon* (see Figure 5.1).

A B-spline can be defined as a linear combination:

$$P(t) = \sum_{i=1}^{n+1} B_i N_{i,r}(t) \quad t_{min} < t < t_{max}, \quad 2 \leq r \leq n+1 \qquad (5.6)$$

$N_{i,r}(t)$ are the normalized basis functions defined by the Cox-de Boor recursion formulas. Specifically

$$N_{i,1}(t) = \begin{cases} 1, & if \ k_i \leq t < k_{i+1} \\ 0, & otherwise \end{cases} \qquad (5.7)$$

and

$$N_{i,r} = \frac{(t - k_i)N_{i,r-1}(t)}{k_{i+r-1} - k_i} + \frac{(k_{i+r} - t)N_{i+1,r-1}(t)}{k_{i+r} - k_{i+1}}. \qquad (5.8)$$

We define

$$\frac{0}{0} = 0.$$

Specific types include the nonperiodic B-spline and uniform B-spline (internal knots are equally spaced). A B-spline with no internal knots is a Bézier curve.

B-spline continuity If the nth derivatives of a curve, $d^n P(t)/dt^n$, at the curve segment joint are equal in both direction and magnitude, then the curve is said to have C^n parametric continuity at the joint.

B-splines automatically take care of continuity, with exactly one control point per curve segment. With different degrees there are many types of B-splines (linear, quadratic, cubic,...) and they may be uniform or non-uniform. We will only consider uniform B-splines for which parametric continuity is always one degree lower than the degree of each curve piece (e.g. linear B-splines have C^0 continuity, cubic have C^2 continuity, etc).

A C^2 curve is doubly differentiable at the knot point and its curvature is continuous.

B-spline curve derivatives. The derivatives of a B-spline curve at any point on the curve are obtained by formal differentiation. Specifically recalling Equation (5.6) the first derivative is

$$P'(t) = \sum_{i=1}^{n+1} B_i N'_{i,r}(t) \tag{5.9}$$

where

$$N'_{i,r}(t) = \frac{N_{i,r-1}(t) + (t - k_i)N'_{i,r-1}(t)}{k_{i+r-1} - k_i} + \frac{(k_{i+r} - t)N'_{i+1,r-1}(t) - N_{i+1,r-1}(t)}{k_{i+r} - k_{i+1}} \tag{5.10}$$

Note from Equation (5.7) that $N'_{i,1}(t) = 0$ for all t.

Consequently, for $r = 2$ Equation (5.9) reduces to:

$$N'_{i,2}(t) = \frac{N_{i,1}(t)}{k_{i+1} - k_i} - \frac{N_{i+1,1}(t)}{k_{i+2} - k_{i+1}}$$

The second derivative is given by:

$$P''(t) = \sum_{i=1}^{n+1} B_i N''_{i,r}(t) \tag{5.11}$$

Differentiating equation (5.10) yields the second derivative of the basis function:

$$N''_{i,r}(t) = \frac{2N'_{i,r-1}(t) + (t - k_i)N''_{i,r-1}(t)}{k_{i+r-1} - k_i} + \frac{(k_{i+r} - t)N''_{i+1,r-1}(t) - 2N'_{i+1,r-1}(t)}{k_{i+r} - k_{i+1}} \tag{5.12}$$

Here, note that both $N''_{i,1}(t) = 0$ and $N''_{i,2}(t) = 0$ for all t. [dB78].

B-spline curve curvature. A plane curve curvature is a geometric property of curve which represents how the curve bends. The plane curve curvature is often defined by following equations

$$K(\psi, s) = \frac{\partial \psi}{\partial s} \tag{5.13}$$

where $k(\psi, s)$ is the curvature, s is an arc length of $p(t) = (x(t), y(t))$, and ψ is an angle between a tangent of $p(t)$ and x-axis. The useful parametric definition is as follows:

$$K(t) = \left| \frac{x''y' - x'y''}{(x'^2 + y'^2)^{\frac{3}{2}}} \right| \tag{5.14}$$

where $K(t)$ is the relative curvature.

For the sake of simplicity of computation in this work, we choose the knot vector

$$K_{\mathrm{m}} = [\frac{l}{2} \ \frac{l}{2} \ \frac{l}{2} \ \frac{l}{2} \ \frac{l}{6} \ -\frac{l}{6} \ -\frac{l}{2} \ -\frac{l}{2} \ -\frac{l}{2} \ -\frac{l}{2}]$$

and it is used to generate a fourth-order (cubic) B-spline curve with a control pentagon.

We define the coordinates of the 6 control points in z direction as

$$B_z = [0.01 \ 0.006 \ 0.002 \ -0.002 \ -0.006 \ -0.01].$$

The B-spline is symmetric to the line $z = 0$ (see Figure 5.1). Therefore, we have three unknown parameters y_0^*, y_1 and y_2^*. Here we consider y_0^* and y_2^* as design variables, as the third unknown parameter (e.g. y_1 in Figure 5.1) can be determined by the mass constraint.

To avoid more strange shapes, we define two constraints.

- $y_0^* \leq y_1 \leq y_2^*$ or
- $y_2^* \leq y_1 \leq y_0^*$.

The value of y_1 should be between y_0^* and y_2^*, thus $y(z)$ is either a convex or a concave function.

Optimization Objectives. Here we define the two objectives in details.

- **Maximum amplitude.** From the computation in Chapters 3 and 4, we know that the computation for the vibration amplitude is complicated and cannot be expressed

69

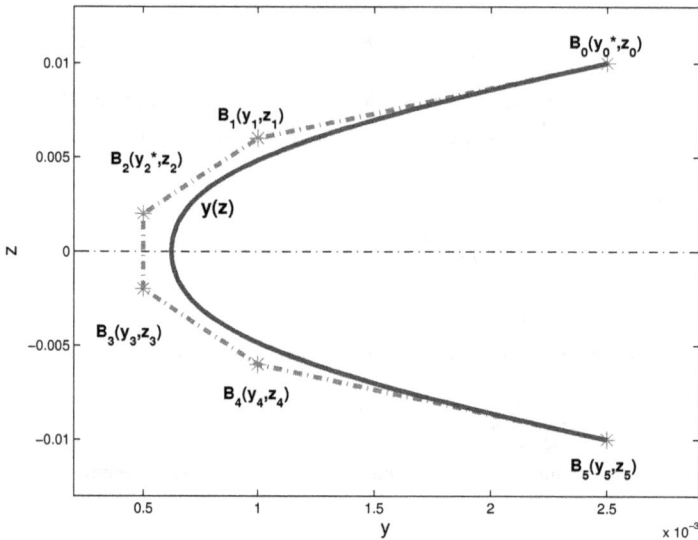

Figure 5.1: A cubic B-spline curve with its control polygon (dash-dot line).

by an explicit formula.

Figure 5.2 illustrates the computation for the vibration amplitude of a piezo step by step. For each input $y^* \in Q$, the following computation steps should be taken:

- compute $y(z)$ by the mass constraint;

- compute numerical solutions of eigenfunctions for a piezo with a curved side described by $y(z)$ by solving a boundary value problem;

- use the obtained numerical values of eigenfunctions to compute the coefficients in equations (4.1) and (3.44);

- call AUTO2000 for computing the vibration amplitude;

- choose the amplitude at a specified excitation frequency.

- **Minimum curvature.** In Figure 3.1, the curved side is described by $y(z)$. For a two-dimensional curve written in the form $y = f(z)$, the equation of curvature

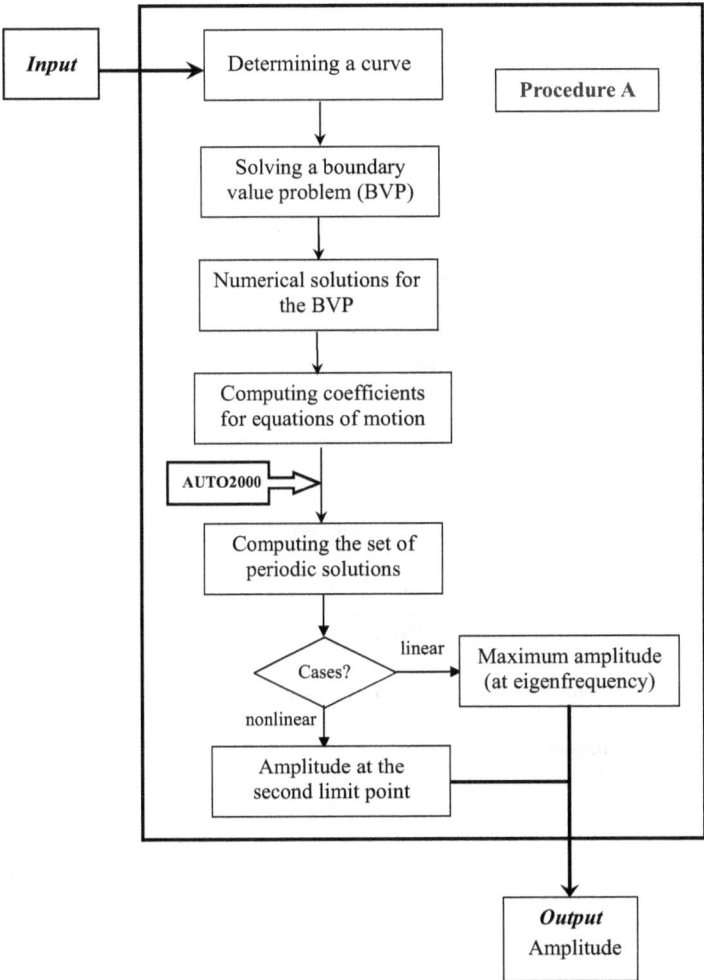

Figure 5.2: Framework for computing amplitude (Procedure A)

becomes

$$K(z) = \frac{f''(z)}{(1 + (f'(z))^2)^{\frac{3}{2}}}.$$ (5.15)

1. $y(z) = az^2 + b$;

   Since

   $$f''(z) = 2a, \quad f'(z) = 2az,$$

   equation (5.15) is simplified to

   $$K(z) = \frac{2a}{(1 + (2az)^2)^{\frac{3}{2}}}.$$

   As the curve $y(z)$ is symmetric to $z = 0$, we choose the absolute value of curvature at $z = 0$ as the other objective

   $$min \quad K = |2a|$$

2. B-spline.

   As discussed in Chapter 3, the relative curvature $K(t)$ for a B-spline at one point is given by:

   $$K(t) = \left| \frac{x''y' - x'y''}{(x'^2 + y'^2)^{\frac{3}{2}}} \right|.$$ (5.16)

Constraints.   The above two objectives are subjected to the following constraints:

- **Mass constraint.** The mass of the piezo is fixed. Only the shape is set to vary. This constraint is expressed by (see Figure 3.1)

$$\int_{-l/2}^{l/2} y(z) dz = const.$$

- **Spatial constraint.** The piezo is symmetric with respect to the y-axis. As shown in Figure 3.1, $y(z)$ is being even ($y(z) = y(-z)$), thus we can restrict the problem to half of the domain $z \in [0, l/2]$.

- **Domain constraint.** The domain Q is defined by

1. $y(z) = az^2 + b$ (one dimension);

$$Q = \{y^* \in ([0.0003, 0.004])\}.$$

2. B-spline (two dimensions).

$$Q = \{(B_0(z), B_2(z)) \in \mathbb{R}^2 | 0.0003 < B_0(z) < 0.004, 0.0003 < B_2(z) < 0.004\}.$$

5.4.3 A GAIO Model for the Optimization Problem

According to the above information, we now introduce a GAIO model to solve the above multi-objective optimization problem (see Figure 5.3).

In principle each of the algorithms proposed in Section 5.3 is applicable to a multi-objective optimization problem on its own. In our case the *subdivision algorithm* is performed.

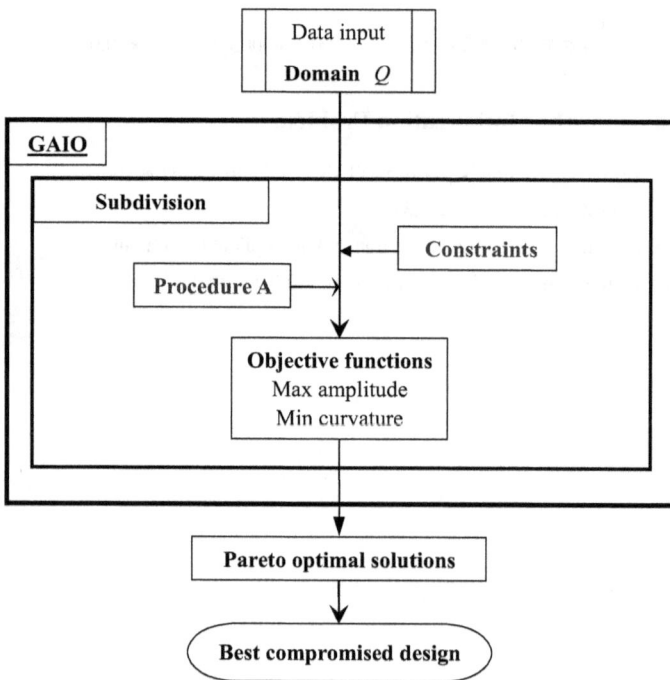

Figure 5.3: A GAIO model for the optimization problem

Chapter 6

Results and Discussion

In Chapter 5, a two-objective shape optimization problem for piezoelectric actuators is formulated, and then a GAIO model is given to find the Pareto sets for the optimization problem. In this chapter, the computation results presented in Section 6.1 are obtained after performing the computation steps in Chapter 5. Problems with one and two design variables are considered respectively in both linear and nonlinear cases. A brief discussion of the results is given in Section 6.2.

6.1 Optimization Results

In this work, we want to design a piezo with one boundary defined by a function $y(z)$, which will be optimized to obtain maximum vibration amplitude and minimum curvature at one time. Performing the computation steps in Figure 5.3, the optimization solutions are presented below in two items. In each item both linear and nonlinear cases are considered.

6.1.1 Multi-Objective Optimization Solutions for One Parameter (quadratic curves)

In Section 3.2.3, a quadratic curve $y(z) = az^2 + b$ was used to represent the boundary $y(z)$ in Figure 3.1, and $y^* = y(0)$ is the design variable.

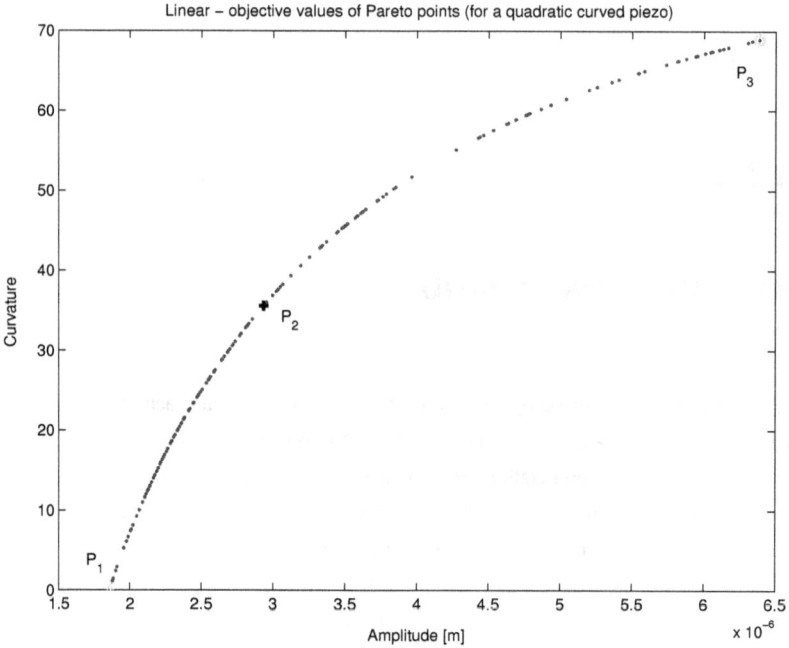

Figure 6.1: Pareto set for two-objective shape optimization problem (linear case, one parameter) .

Linear case.   Figures 6.1 and 6.2 are the Pareto set and its preimages in the linear case respectively. Two objectives are obtained for maximum amplitude and minimum curvature. The maximum amplitude is calculated by performing Procedure A in Figure 5.2. The other objective curvature is obtained at $z = 0$ as $K = |2a|$.

The Pareto solutions in Figure 6.1 are well collocated as a smooth curve. The preimages in Figure 6.2 are in the range of $y^* \in [0.00035, 0.0015]$, that is, the corresponding shapes are all concave. Example shapes of points P_1, P_2 and P_3 are also given in Figure 6.2.

76

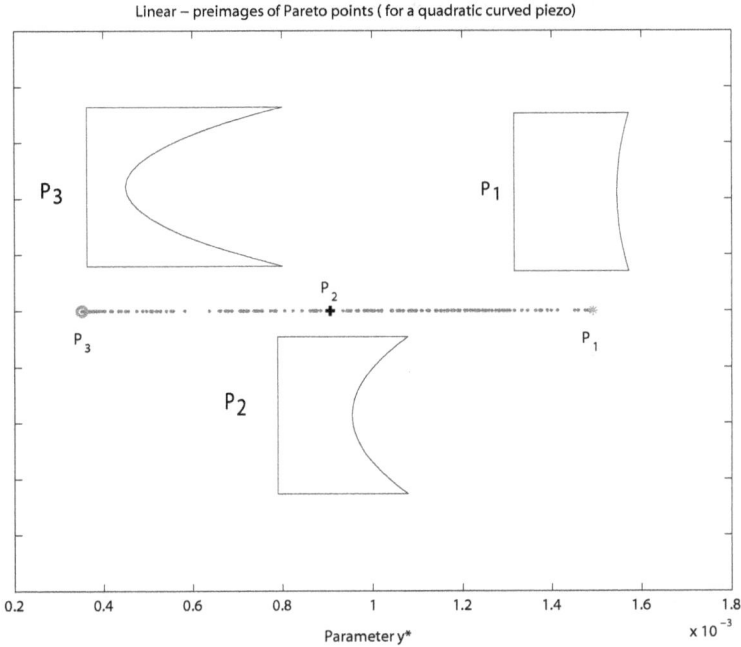

Linear – preimages of Pareto points (for a quadratic curved piezo)

Figure 6.2: Preimages of Pareto set for two-objective shape optimization problem (linear case, one parameter) .

Nonlinear Case. In Figures 6.3 and 6.4 the Pareto set and their preimages in the non-linear case are plotted respectively.

Two objectives are also obtained for maximum amplitude and minimum curvature. The objective curvature is obtained in the same way as in the linear case. Because of the bifurcation in the nonlinear case there are unstable solutions as well as stable ones. Then the objective amplitude is focused on the amplitude at the right limit point, that is, the maximum amplitude of the stable solutions (as shown in Figure 4.4).

In Figure 6.3 we observed a "jump phenomenon", which separates the Pareto solutions

77

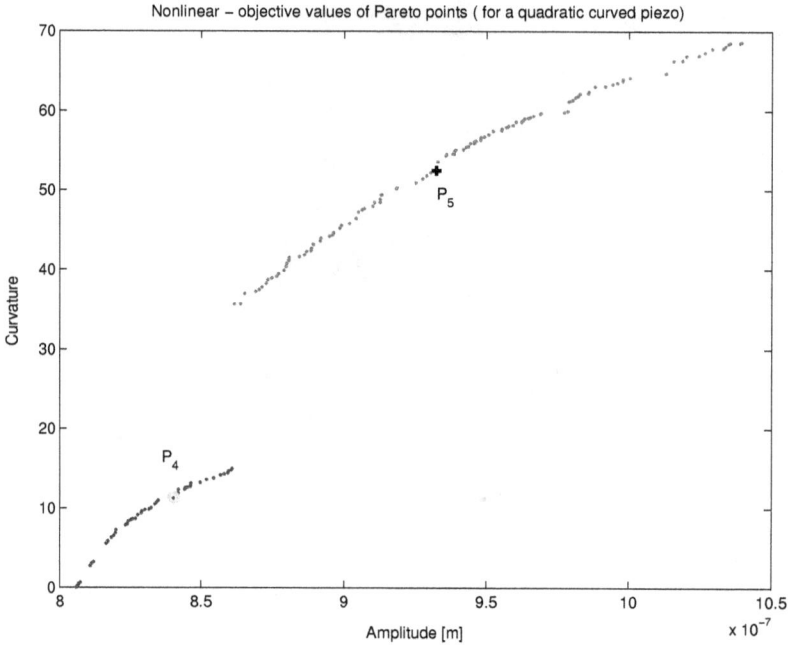

Figure 6.3: Pareto set for two-objective shape optimization problem (nonlinear case, one parameter).

into two sets. Looking at their preimages in Figure 6.4, we found that the preimages y^* are also in two sets in range of $[0.00035, 0.00091]$ and $[0.0015, 0.00175]$, and their corresponding shapes are concave and convex respectively. Example shapes of points P_4, and P_5 are also given in Figure 6.4.

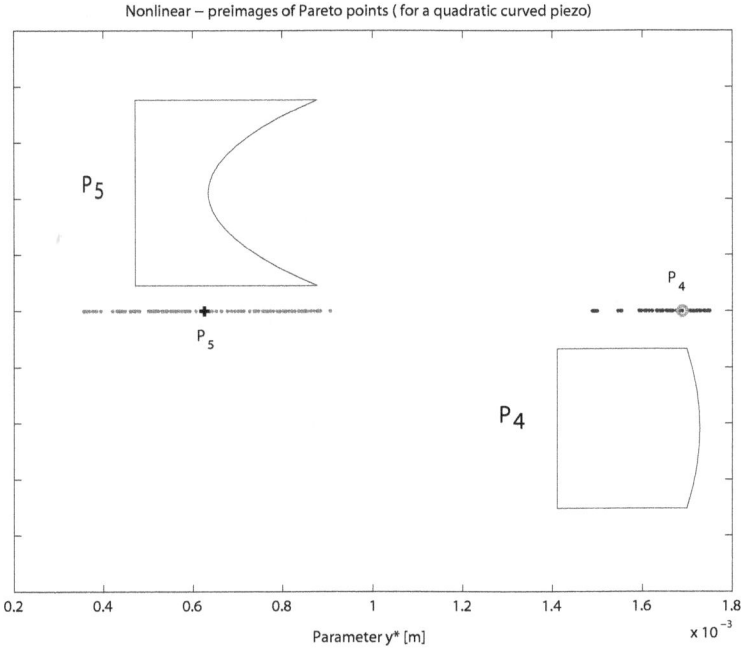

Figure 6.4: Preimages of Pareto set for two-objective shape optimization problem (nonlinear case, one parameter).

6.1.2 Multi-Objective Optimization Solutions for Two Parameters (cubic B-spline curves)

In Section 3.2.3, we also introduced another way to represent a curve with two design variables. It is a cubic B-spline curve with six control points (see Figure 5.1). y_0^* and y_2^* in Figure 5.1 are the design variables.

Linear case. Figures 6.5 and 6.6 are the Pareto set and their preimages in the linear case respectively. Two objectives are obtained in the same way as in one design variable case

79

in Section 6.1.1.

In Figure 6.5, we also observed a "jump phenomenon". When looking at their preimages in Figure 6.6, the preimages are in one set in range of $y_0^* \in [0.0027, 0.003]$, $y_2^* \in [0.0007, 0.001]$. Example shapes of points P_6 and P_7 are plotted in Figure 6.6.

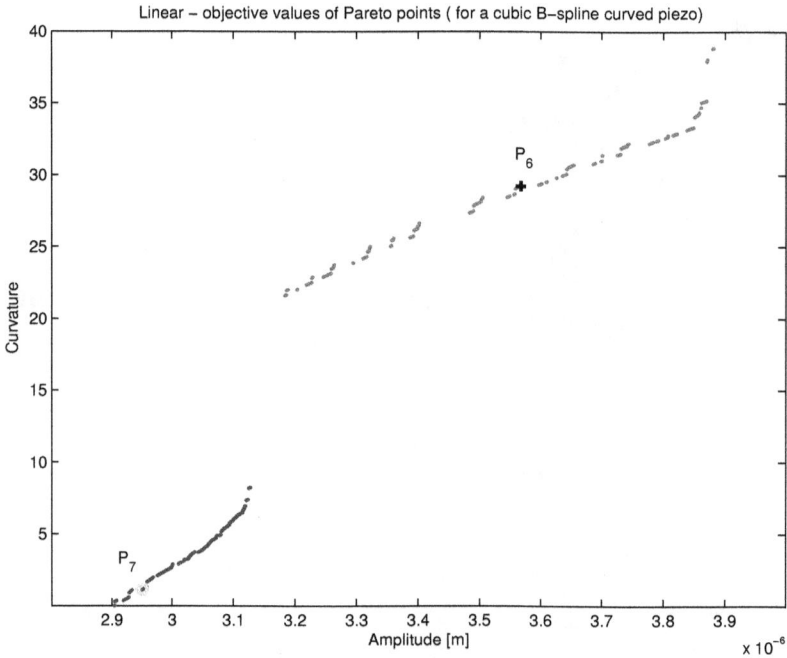

Figure 6.5: Pareto set for two-objective shape optimization problem (linear case, two parameters).

Nonlinear case.   For the nonlinear case, the Pareto points and their preimages are plotted in Figures 6.7 and 6.8 respectively. In Figure 6.8 the preimages are mostly located within the range of $y_0^* \in [0.00165, 0.0025]$, $y_2^* \in [0.00085, 0.00135]$, only one point P_8 is at

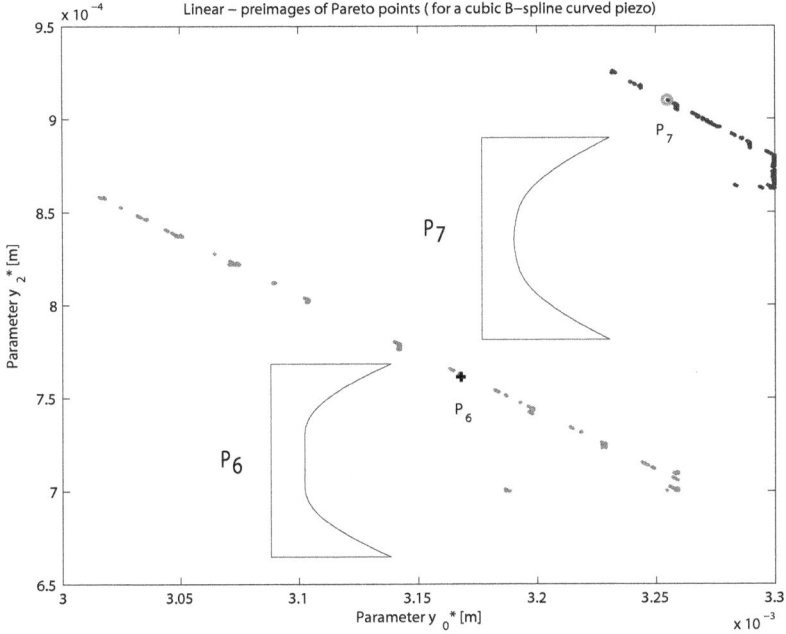

Figure 6.6: Preimages of Pareto set for two-objective shape optimization problem (linear case, two parameters).

(0.00137,0.00155). It means that most Pareto shapes are concave, and one is convex. Example shapes of points P_8 and P_9 are plotted in Figure 6.8.

6.2 Discussion

In this section, we will discuss the results of Section 6.1.

The optimization results with one and two parameters for a linear model are compared in Figure 6.9. Two example shapes, P_{10} and P_{11}, are plotted in Figure 6.9. It is obvious that results with two parameters are better than those with one parameter. Concretely, the

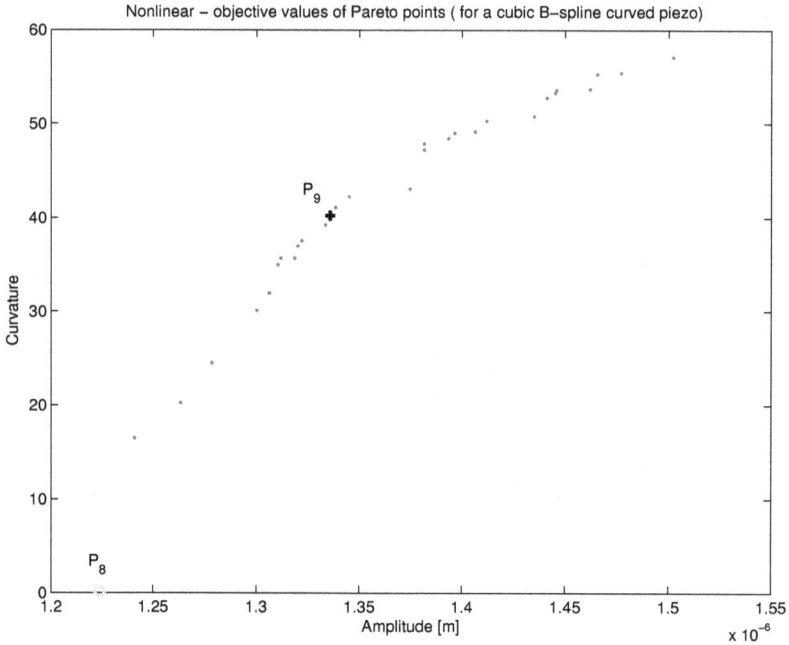

Figure 6.7: Pareto set for two-objective shape optimization problem (nonlinear case, two parameters).

curvatures at points P_{10} and P_{11} are 29.9 and 30.0 respectively, but the amplitude at P_{11} is $3.6372 \cdot 10^{-6}$, which is 35.43% higher than the amplitude $2.6857 \cdot 10^{-6}$ at point P_{10}.

The optimization results for one and two parameters in the nonlinear case are compared in Figure 6.10. Example shapes of Points P_{12} and P_{13} are also plotted. At point P_{12}, the amplitude is $8.7971 \cdot 10^{-7}$, and the curvature is 40.4. At point P_{13}, the amplitude is $1.3358 \cdot 10^{-6}$, and the curvature is 40.2. Comparing the objective values of points P_{12} and P_{13}, the curvature of P_{13} is almost the same as that of P_{12}, but the amplitude is improved by 51.85%.

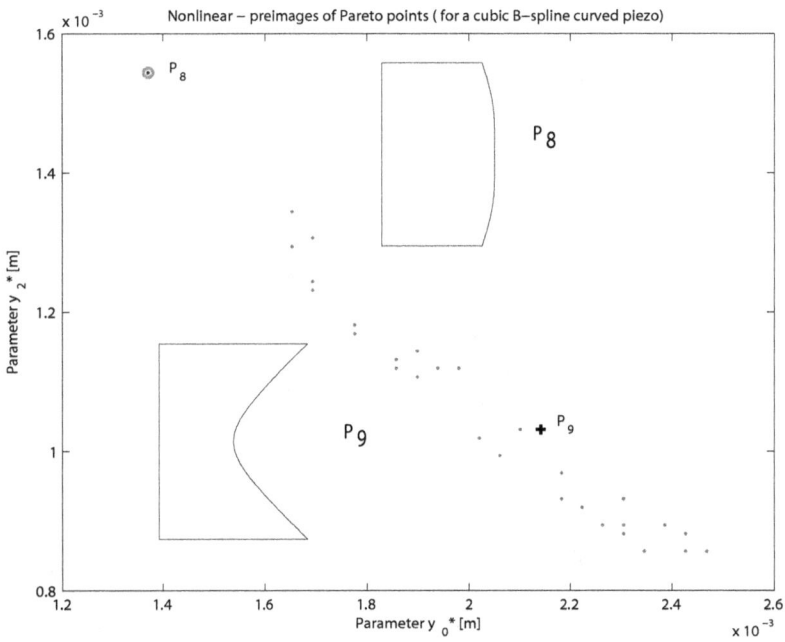

Figure 6.8: Preimages of Pareto set for two-objective shape optimization problem (nonlinear case, two parameters).

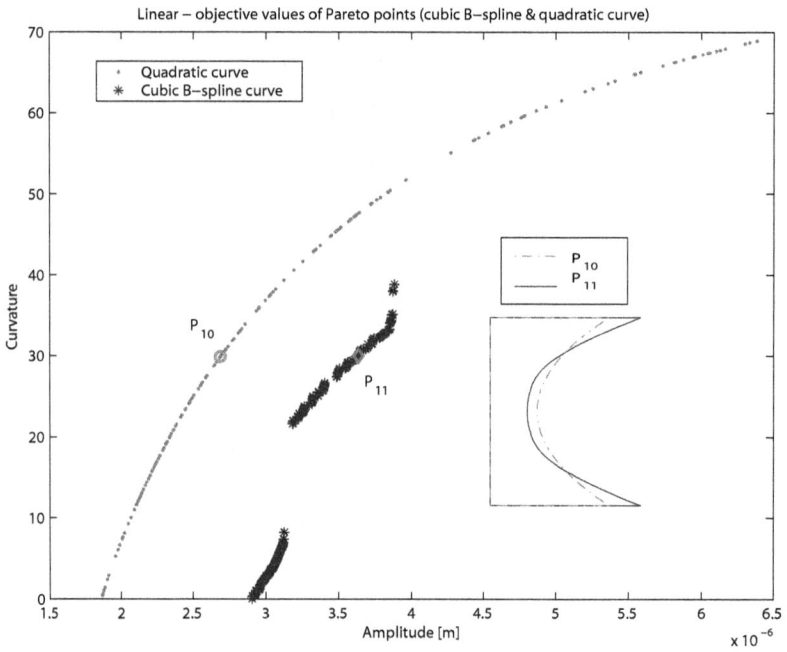

Figure 6.9: Pareto sets for two-objective shape optimization problem (linear case,).

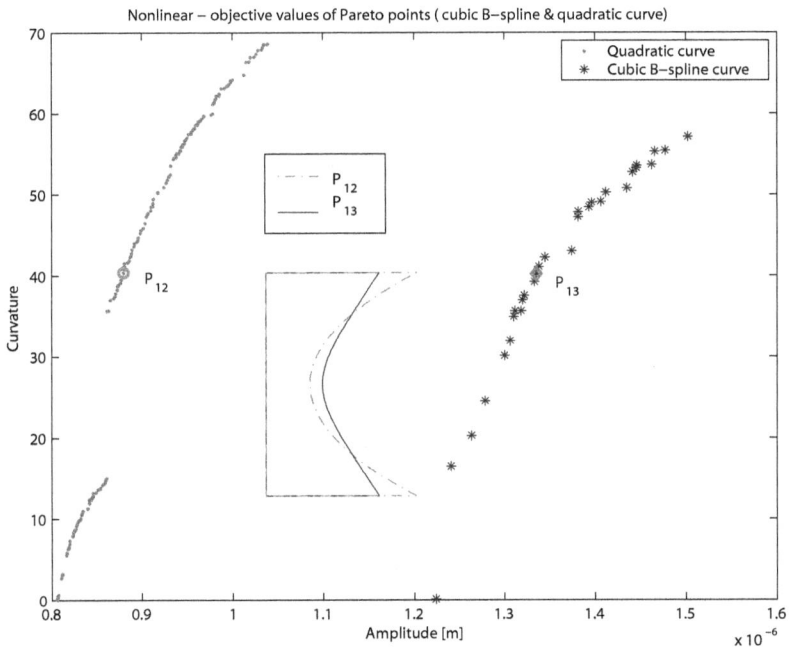

Figure 6.10: Pareto sets for two-objective shape optimization problem (nonlinear case).

6 Results and Discussion

Chapter 7

Summary and Outlook

In this chapter a summary of the results of this dissertation is given and future research is envisaged and recommended.

7.1   Summary

Shape optimization is now a major concern in the design of mechanical systems in industry. It is important to improve the performance of a piezo by changing its geometry. However, academic and industrial research into shape optimization is still ongoing. Motivated by facts that a curved side piezo has a performs better than a rectangular sided piezo in calculations performed by using a simulating software, and convinced that this information could facilitate the shape optimization of the piezoceramics with respect to several objectives, the numerical analysis, modeling and multi-objective optimization of the shape of piezoceramics have been addressed. This dissertation is comprised of three phases:

1. Background,

2. modeling and numerical analysis,

3. multi-objective shape optimization.

In the first phase, the basic knowledge of shape optimization, piezoelectric effects, piezoelectric materials and their properties were introduced. The influence of the shape of a piezo on its performance was demonstrated with software package Comsol Multiphsics (FEMLAB). The preliminary results were visualized in Figures 1.2 and 1.3. The motivation and the primary goal of this work have been also given.

In the second phase, a detailed mathematical model able to reproduce the dynamical behavior (in particular some nonlinear phenomenon) of piezoceramics is first introduced. Then, the general eigenfunctions of piezoceramics for different geometries, which describe how a shape(geometry) of a piezo influences its properties, are derived via Hamilton's principle. Both usual (e.g. rectangular) and unusual (e.g. curved side) shapes are considered. A corresponding boundary value problem is solved numerically using MATLAB and the computation results are used to compute the coefficients of the equation of motion.

Then the linear equation of motion of piezoceramics for different geometries is derived via the Calculus of Variations. Compared with the linear equation of motion, two nonlinear terms are introduced. The equation of motion is solved numerically via a continuation software package AUTO2000, and the results show that some curved side piezoceramics perform better than those with a rectangular shape for both linear and nonlinear models. The difference between the solutions of linear and nonlinear cases is that there are unstable solutions and two limit points in the nonlinear case. The effects of the two nonlinear terms are analyzed, and both nonlinear terms have profound effects on the bifurcation observed.

Finally, a multi-objective shape optimization problem for the design of piezoelectric actuators is introduced in the third phase. Two objectives are maximum amplitude (better performance) and minimum curvature (simple manufacturing). The framework for computing the amplitude is established. A GAIO model for the optimization problem is proposed. The optimization is conducted with subdivision algorithm based on GAIO software package, and the corresponding Pareto-optimal solutions are obtained. Results show that the Pareto set with two design variables (e.g. a cubic B-spline curved side shape) is better than the Pareto set with one design variable (e.g. a quadratic curved side shape) for both linear and nonlinear models.

7.2 Outlook

Although some useful results are obtained, there is still much work should be done in the futher. The work could be further expanded in a number of ways to enhance its capability for supporting industrial practices. I specifically recommend the followings:

1. One of the limitations of this work is the reduction of the number of design variables. I considered the cases of one and two design variables, respectively. More variables could be beneficial.

2. A two-dimensional problem is considered and represented by quadratic curves and cubic B-spline curves. A three-dimensional problem which would be represented by B-spline surfaces maybe worth investigating in the future.

3. Two objectives (max. amplitude & min. curvature) have been considered in this work. In the design of piezoelectric actuators, more objectives (e.g. min. input electric field) or maybe minimum amplitude in some cases need to be optimized to meet the design requirements. It is recommended that more than two objectives will considered in the future in order to improve feasibility and capability.

4. The work presented so far tackles some simplified cases to improve the performance of a piezo. Experiments are welcome to verify simulation results.

7 Summary and Outlook

Appendix A

COMSOL Multiphysics (former FEMLAB)

COMSOL Multiphysics (former FEMLAB) is an interactive environment for modeling and simulating scientific and engineering problems based on partial differential equations (PDEs)-equations that are the fundamental basis for the laws of science. COMSOL is a package that is based off of Matlab and is a contraction for Finite Element Method Laboratory. The Finite Element Method, or FEM for short, is a numerical method that can be used to solve PDEs.

COMSOL Multiphysics is a complete package that covers all facets of the modeling process. It contains CAD tools, interfaces for physics and equation specifications, automatic mesh generation, a variety of optimized solvers, as well as visualization and postprocessing tools. Its multiphysics capability allows to simultaneously modeling many combinations of coupled phenomena and allows us to supplement ready-to-use applications based on predefined relevant physical quantities with equation-based modeling.

It gives here a quick overview of some of the features available for advanced modeling in COMSOL's graphical user interface. The important COMSOL features are:

- Fast, interactive, and user-friendly graphical user interface for all steps of the modeling process;

- Powerful direct and iterative solvers;

- Linear and nonlinear stationary, timedependent, and eigenvalue analysis of models;

- Total freedom in the specification of physical properties, whether as analytical expressions or functions;

- Unlimited multiphysics capabilities for coupling all types of physics, even on domains in different space dimensions;

- General formulations for quick and easy modeling of arbitrary systems of PDEs;

- CAD tools for solid modeling in 1D, 2D and 3D;

- Triangular, quadrilateral, tetrahedral, brick, and prism meshes using fully automatic and adaptive mesh generation;

- Extensive model libraries that document and demonstrate more than 100 solved examples;

- Parametric solver for efficient solution of highly nonlinear models;

- Interactive postprocessing and visualization;

- Report generator for documenting models;

- 64-bit platform support for large-scale computations;

- Smooth interface to MATLAB.

Using COMSOL Multiphysics.   With COMSOL Multiphysics' interactive modeling environment you can build and analyze models from start to finish without the need to involve any other software packages. Its integrated tools allow you to work efficiently at each step in the process, all within one consistent and easy-to-use graphical environment. It's easy to move back and forth between various stages such as setting up the geometry, defining the physics, creating a mesh, solving the model, and performing postprocessing. COMSOL Multiphysics' associative geometry feature preserves any boundary condition or equation even if you change the geometry. The modeling procedure typically involves the following steps:

1. **Create of the geometry;**

   COMSOL Multiphysics provides powerful CAD tools for creating 1D, 2D and 3D geometric objects using solid modeling. Work planes are useful for generating 2D profiles that you rotate, extrude, and embed into 3D structures.

2. **Define the physic;**

   COMSOL Multiphysics makes the modeling of many physical processes and equations effortless through a variety of predefined application modes.

3. **Generate the Finite Element Mesh;**

   Built-in mesh generators automatically perform meshing. They can create triangular or tetrahedral unstructured meshes as well as quadrilateral meshes. By extruding or revolving a 2D mesh, you can create brick and prism meshes.

4. **Compute the Solution;**

   COMSOL Multiphysics runs time-dependent or stationary simulations for linear and nonlinear systems. With its solver scripting language, one can manage and automate the solution process to solve for different field variables or iterate using a staged-solution approach.

5. **Visualize and Postprocess the Results;**

6. **Perform Optimization and Parametric Analysis.**

   The parametric solver in COMSOL Multiphysics provides the perfect way for examining a series of conditions. In addition, the built-in MATLAB interface can save COMSOL Multiphysics models as M-files for later incorporation as functions into MATLAB scripts for optimization or other postprocessing.

More information see:

http://www.comsol.com.

A   COMSOL Multiphysics (former FEMLAB)

Appendix B

A Brief Introduction to AUTO2000

AUTO is a publicly available software for continuation and bifurcation problems in ordinary differential equations developed by Eusebius Doedel. It was originally written in 1980 and widely used in the dynamical systems community.

AUTO can do a limited bifurcation analysis of algebraic systems of the form

$$f(u,p) = 0, \quad f, u \ in \ R^n$$

and of systems of ordinary differential equations of the form

$$u'(t) = f(u(t),p), \quad f, u \ in \ R^n$$

subject to initial conditions, boundary conditions, and integral constraints. Here p denotes one or more parameters. AUTO can also do certain continuation and evolution computations for parabolic PDEs. It also includes the software HOMCONT for the bifurcation analysis of homoclinic orbits.

In AUTO the computation of periodic solutions to a periodically forced system can be done by adding a nonlinear oscillator with the desired periodic forcing as one of the solution components ([ADO90]).

An example of such an oscillator is

$$\begin{aligned} x' &= x + \beta y - x(x^2 + y^2), \\ y' &= -\beta x + y - y(x^2 + y^2), \end{aligned}$$

95

which has the asymptotically stable solution

$$x = \sin(\beta t), \ \ y = \cos(\beta t)$$

Coupling this oscillator to the Fitzhugh-Nagumo equations:

$$
\begin{aligned}
v' &= (F(v) - w)/\varepsilon, \\
w' &= v - dw - (b + r\sin(\beta t)),
\end{aligned}
$$

by replacing $\sin(\beta t)$ by x. Above, $F(v) = v(v-a)(1-v)$ and a, b, ε and d are fixed. The first run is a homotopy from $r = 0$, where a solution is known analytically, to $r = c$, where c is a positive constant. Part of the solution branch with $r = c$ and varying β is computed in the second run. β is treated as the bifurcation parameter.

AUTO2000 is freely available from https://sourceforge.net/projects/auto2000/.

Bibliography

[ADO90]   J. C. Alexander, E. J. Doedel, and H. G. Othmer. On the resonance structure in a forced excitable system. *SIAM J. APPL. MATH.*, 50(5):1373–1418, 1990.

[AJT02]   G. Allaire, F. Jouve, and A. Toarder. A level-set method for shape optimization. *C. R. Acad. Sci. Paris*, Série 1:1–6, 2002.

[Arn78]   V. I. Arnold. *Mathematical methods of classical mechanics*. Springer-verlag, New York, 1978.

[BF04]   S. Bharti and M. I. Frecker. Optimal design and experimental characterization of a compliant mechanism piezoelectric actuator for inertially stabilized rifle. *Journal of Intelligent Material Systems and Structures*, 15(2):93–106, 2004.

[BK88]   M. Bensdøe and N. Kikuchi. Generating optimal topologies in structural design using a homogenization method. *Computer Methods in Applied Mechanics and Engineering*, 71(2):197–224, 1988.

[BLTZ03]   S. Bleuler, M. Laumanns, L. Thiele, and E. Zitzler. PISA - a platform and programming language independent interface for search algorithms. In *Conference on Evolutionary Multi-criterion optimization (EMO 2003)*, pages 494–508, Portugal, 2003.

[BRG03]   P. Bürmann, A. Raman, and S. V. Garimella. Dynamics and topology optimization of piezoelectric fans. *IEEE Transactions on Components and Packaging Technologies*, 25(4):592–600, 2003.

[BTO97]   A. Benjeddou, M. A. Trindade, and R. Ohayon. A unified beam finite element model for extension and shear piezoelectric actuation mechanisms. *Journal of Intelligent Material Systems and Structures*, 8(12):1012–1025, 1997.

[BTO99]   A. Benjeddou, M. A. Trindade, and R. Ohayon. New shear actuated smart structure beam finite element. *AIAA Journal*, 37(3):378–383, 1999.

[CC05]   C. A. Coello Coello. *Recent Trends in Evolutionary Multiobjective Optimization*, volume in A. Abraham, L. Jain and R. Goldberg (editors), Evolutionary Multiobjective Optimization: Theoretical Advances and Applications. 2005.

[CF00]   S. Canfield and M. Frecker. Topology optimization of compliant mechanical amplifiers for piezoelectric actuators. *Structural and Multidisciplinary Optimization*, 20(4):269–279, 2000.

[Che00]   A. Cherkaev. *Variational Methods for Structural Optimization*. Springer-Verlag, 2000.

[dB78]   C. de Boor. *A practical guide to splines*. Springer-Verlab, New York, 1978.

[Dea01]   E. Doedel and R. Paffenroth et al. Auto 2000: Continuation and bifurcation software for ordinary differential equations (with homcont). Technical report, Caltech, 2001.

[Dea02]   K. Deb and A. Pratap et al. A fast and elitist multi-objective genetic algorithm: NSGA-II. *IEEE Transaction on Evolutionary Computation*, 6(2):181–197, 2002.

[Deb]   http://www.iitk.ac.in/kangal/codes.shtml.

[Deb01]   K. Deb. *Multi-Objective Optimization Using Evolutionary Algorithms*. John Wiley & Sons, Chichester, 2001.

[DFJ01]   M. Dellnitz, G. Froyland, and O. Junge. *The algorithms behind GAIO - set oriented numerical methods for dynamical systems*, volume In B. Fiedler (ed.) Ergodic Theory, Analysis, and Efficient Simulation of dynamical Systems. Springer, 2001.

[DH97]   M. Dellnitz and A. Hohmann. A subdivision algorithm for the computation of unstable manifolds and global attractors. *Numerische Mathematik*, 75:293–317, 1997.

[DJW05]   M. Dellnitz, O. Junge, and F. Wang. Exploring the dynamics of nonlinear models for a piezoceramic. In *D. H. van Campen (ed.), Proceedings of ENOC 2005: Fifth EUROMECH Nonlinear Dynamics Conference*, Eindhoven, The Netherlands, 2005.

[DSH05]   M. Dellnitz, O. Schütze, and T. Hestermeyer. Covering pareto sets by multilevel subdivision techniques. *Journal of Optimization Theory and Applications*, 124(1):113–136, 2005.

[DSS02]   M. Dellnitz, O. Schütze, and S. Sertl. Finding zeros by multilevel subdivision techniques. *IMA Journal of Numerical Analysis*, 22(2):167–185, 2002.

[Ehr02]   M. Ehrgott. *Multiple Criteria Optimization: State of the Art Annotated Bibliographic Surveys*. Kluwer Academic Publishers, Secarcus, USA, 2002.

[FF95]   C. M. Fonseca and P. J. Fleming. An overview of evolutionary algorithms in multiobjective optimization. *Evolutionary Computation*, 3(1):1–16, 1995.

[Fu05]   B. Fu. *Piezoelectric actuator design via multiobjective optimization methods*. PhD thesis, University of Paderborn, Paderborn, 2005.

[HM03]   J. Haslinger and R. A. E. Mäkinen. *Introduction to shape optimization: theory, approximation and computation (Advances in design and control)*. SIAM, Philadelphia, 2003.

[HMN05]   E. Heikkola, K. Miettinen, and P. Nieminen. Applying IND-NIMBUS to a design problem in high-power ultrasonics, reports of the department of mathematical information technology. Scientific computing, University of Jyväskylä, Jyväskylä, 2005.

[Hor97]   J. Horn. *Multicriteria Decision Making*, volume in T. Baack, D. B. Fogel and Z. Michalewicz (ed.). Handbook of Evolutionary Computation. Institute of Physics Publishing, Bristo, UK, 1997.

[Ike90]   T. Ikeda. *Fundametals of Piezoelectricity*. Oxford University Press, USA, 1990.

[Jas]   http://www-idss.cs.put.poznan.pl/~jaszkiewicz/momhlib.

[JNL00]   T. Y. Jiang, T. Y. Ng, and K.. Y. Lam. Optimization of a piezoelectric ceramic actuator. *Sensors and Sctuators*, 84:81–94, 2000.

[KC00]   J. D. Knowles and D. W. Corne. Approximating the nondominated front using the pareto archived evolution strategy. *Evolutionary Computation*, 8(2):149–172, 2000.

[LXKS01]   Y. Li, X. Xin, N. Kikuchi, and K. Saitou. Optimal shape and location of piezoelectric materials for topology optimization of flextensional actuators. In *Proceeding of 2001 Genetic and Evolutionary Computation Conference*, USA, 2001.

[MGH94]   J. A. Main, E. Garcia, and D. Howard. Optimal placement and sizing of paired piezoactuators in beams and plates. *Smart Materials and structures*, 3:373–381, 1994.

[Mie99]   K. Miettinen. *Nonlinear Multiobjective Optimization*. Kluwer Academic Publishers, Bosten, 1999.

[MM00]   K. Miettinen and M. M. Mäkelä. Interactive multiobjective optimization system WWW-NIMBUS on the internet. *Computers & Operations Research*, 27:709–723, 2000.

[Mor03]   T. Morita. Miniature piezoelectric motors. *Sensors and Actuators A*, 103(3):291–300, 2003.

[Neu02]   N. Neumann. *Nichtlineare Effekte bei Längsschwingungen axial polarisierter piezokeramischer Stäbe: Experimentelle Untersuchungen und Parameteridentifikation.* 2002. Diplomarbeit.

[PC04]   G. T. Pulido and C. A. Coello Coello. Using clustering techniques to improve the performance of a multi-objective particle swarm optimizer. In *k. Deb, et al (ed.), Proceedings of GECCO 2004: Genetic and Evolutionary Computation Conference*, Seattle, USA, 2004.

[Pie01]   V. Piefort. *Finite Element Modellig of Piezoelectric Active Structures.* PhD thesis, Université Libre de Bruxelles, 2001.

[Pra74]   W. Prager. *Introduction to sturctural Optimization.* Springer-Verlag, Wien, 1974.

[Ric95]   R. A. Richards. *Zeroth-order Shape Optimization Utilizing a Learning Classifier System.* PhD thesis, Stanford University, 1995.

[Rog01]   D. F. Rogers. *An Introduction to NURBS, With Historical Perspective.* Morgan Kaufmann Publishers, 2001.

[Sch04]   O. Schütze. *Set Oriented Methods for Global Optimization.* PhD thesis, Universität Paderborn, 2004.

[Sea05]   M. K. Samal and P. Seshu et al. A finite element model for nonlinear behaviour of piezoceramics under weak electric fields. *Finite Elements in Analysis and Design*, 41:1464–1480, 2005.

[SZ92]   J. Sokolowski and J. Zolesio. *Introduction to Shape Optimization: Shape Sensitivity Analysis.* Springer-Verlag, New York, 1992.

[Tea01]   K. C. Tan and T. H. Lee et al. A multi-objective evolutionary algorithm toolbox for computer-aided multi-objective optimization. *IEEE Transactions on Systems, Man and Cybernetics: Part B (Cybernetics)*, 31(4):537–556, 2001.

101

[Tho02]   M. L.. Thompson. *On the Material Properties and Constitutive Equations of Piezoelectric Ploy Vinylidene Fluoride(PVDF)*. PhD thesis, Drexel University, 2002.

[vVL00]   D. A. van Veldhuizen and G. B. Lamont. Multiobjective evolutionary algorithms: Analyzing the state-of-the-art. *Evolutionary Computation*, 8(2):125–147, 2000.

[vWH02]   U. von Wagner and P. Hagedorn. Piezo-beam systems subjected to weak electric field: Experiments and modeling of nonlinearities. *Journal of Sound and Vibration*, 256(5):861–872, 2002.

[vWH03]   U. von Wagner and P. Hagedorn. Nonlinear effects of piezoceramics excited by weak electric fields. *Nonlinear Dynamics*, 31:133–149, 2003.

[Wer02]   P. J. Werbos. Classical ODE and PDE which obey quantum dynamics. *International Journal of Bifurcation and Chaos*, 12(10):2031–2049, 2002.

[Zit99]   E. Zitzler. *Evolutionary Algorithms for Multiobjective Optimization: Methods and Applications*. PhD thesis, Swiss Federal Institute of Technology (EM) Zurich, 1999.

List of Symbols

$()'$	$\frac{\partial}{\partial z}$
$\dot{()}$	time derivation
a	constant
A	area of cross-section
b	constant
B_n	control points
$\mathcal{B}_k$	box
α	electromechanical transformation factor
C	parametric continuity
d	degree of B-spline
d, d^t	piezoelectric coupling
d_{33}	piezoelectric 33-effect
D	charge density
$\mathcal{D}$	neighborhood
E^0	Young's modulus of piezoceramic material
E^1, E^2	purely mechanical nonlinear parameter
E, E_z	electric field
$F(x)$	objective function

γ_0	linear electromechanical parameter
γ_1^1, γ_1^2	electromechanical quadratic nonlinear parameter
$\gamma_2^1, \gamma_2^2, \gamma_2^3$	electromechanical cubic nonlinear parameter
$g(x)$	inequality constraint
$h(x)$	equality constraint
H	electric enthalpy density
J	functional
k	$1, 2, \ldots$
k	number of objective functions
K	curvature
K_{m}	knot vector
λ_k	eigenvalue
l	length of piezoceramic
L	Lagrangian function
L	linear differential operator
m	number of inequality constraints
m	number of knots
n	number of decision variables
$N_{i,x}(t)$	basis functions of B-spline
n	number of control points
ν_0	linear purely electrical parameter
ν_1, ν_2	purely electrical nonlinear parameter
$\epsilon, \epsilon^T, \epsilon_{33}^T$	permittivity
ω_k	eigenfrequency
Ω	excitation frequency
$\varphi(z,t)$	electric potential

104

$\Phi(z)$	eigenfunction for electric potential
ψ	angle between a tangent of a plane curve and x-axis.
p	number of equality constraints
$P(t)$	B-spline
Q	domain
r	order of B-spline
ρ	density of piezoceramic
s, s_E	mechanical compliance
S, S_{zz}	strain
$\mathcal{S}$	set
S	shape of a piezo
t, t_0, t_1	time
T	stress
T	kinetic energy density
U_0	amplitude of excitation voltage
$w(z, t)$	displacement at any point along z-axis
$W_k(z)$	eigenfunction for mechanical displacement
$\delta \mathsf{W}$	virtual work of external mechanical and electrical forces
x	vector of decision variables
$\mathcal{X}$	set
$y(z)$	piezo shape function
y^*	design variable
z	z-axis

LIST OF SYMBOLS

List of Tables

1.1   Material properties   8

1.2   Beam's tip deflections for example shapes of piezoceramics   10

2.1   Applications of piezoelectric actuators   23

5.1   Some software packages for MOOPs   61

LIST OF TABLES

List of Figures

1.1 Example: bending of a beam 7

1.2 A static 3D example with Comsol Multiphsics (FEMLAB): bending of a beam (with a cuboid piezo) 9

1.3 A static 3D example with Comsol Multiphsics (FEMLAB): bending of a beam (with a curved surface piezo) 11

2.1 schematic diagram of piezoelectric effects 16

2.2 Strain change associated with the polarization reorientation (adapted from [Pie01]) 19

2.3 Piezoelectric elementary cell (a) before poling (b) after poling 21

3.1 Shape of a piezo under consideration. 27

3.2 Eigenfunctions of the piezo for $y(z) = a|z| + b$ 39

3.3 Eigenfunctions of the piezo for $y(z) = az^2 + b$ 40

3.4 Plotting of $w(z,t) = \sum_{i=1}^{k} W_i(z)p_i(t)$, $k = 1, 4$ respectively 45

3.5 Plotting of $W_k(z)p_k(t)$, $k = 1, 2, 3, 4$ respectively 46

4.1 The periodic solutions of the linear equation of motion for the rectangular piezo 49

4.2 Linear model: oscillation amplitude of the piezo dependent on the excitation frequency for different geometries: rectangular (green) and curved shape (blue). 50

4.3 Nonlinear model: path of periodic solutions within a certain range of excitation frequencies (piezo with rectangular shape). 52

4.4   Nonlinear model: paths of periodic solutions within a certain range of excitation frequencies for different geometries of the piezo: rectangular (green) and curved shapes (blue). 53

4.5   Nonlinear model: paths of periodic solutions within a certain range of excitation frequencies for different values of the parameter ε. 55

4.6   Nonlinear model: paths of periodic solutions within a certain range of excitation frequencies for different values of the parameter ε_d. 56

5.1   A cubic B-spline curve with its control polygon (dash-dot line). 70

5.2   Framework for computing amplitude (Procedure A) 71

5.3   A GAIO model for the optimization problem 74

6.1   Pareto set for two-objective shape optimization problem (linear case, one parameter) 76

6.2   Preimages of Pareto set for two-objective shape optimization problem (linear case, one parameter) 77

6.3   Pareto set for two-objective shape optimization problem (nonlinear case, one parameter). 78

6.4   Preimages of Pareto set for two-objective shape optimization problem (nonlinear case, one parameter). 79

6.5   Pareto set for two-objective shape optimization problem (linear case, two parameters). 80

6.6   Preimages of Pareto set for two-objective shape optimization problem (linear case, two parameters). 81

6.7   Pareto set for two-objective shape optimization problem (nonlinear case, two parameters). 82

6.8   Preimages of Pareto set for two-objective shape optimization problem (nonlinear case, two parameters). 83

6.9   Pareto sets for two-objective shape optimization problem (linear case,). . 84

6.10 Pareto sets for two-objective shape optimization problem (nonlinear case). 85

www.ingramcontent.com/pod-product-compliance
Lightning Source LLC
Chambersburg PA
CBHW070738220326
41598CB00024BA/3470